I0760284

Arithmetic Island

The Cursed Dragon

Written & Illustrated by: Kyle Whiting

For permissions, inquiries, or more information, visit:

kylewhiting.com

ISBN: 978-1-967360-04-8

This is a work of fiction. Names, characters, places, and events are either the product of the author's imagination or used fictitiously. Any resemblance to actual persons, living or dead, or real events is purely coincidental.

Published by Kyle Whiting

Printed in the United States of America

Cover design by Kyle Whiting

Illustrations by Kyle Whiting

Disclaimer

This is a work of fiction. All characters, events, and settings are products of the author's imagination or are used fictitiously. The story includes references to historical figures such as Pythagoras, Hypatia, Al-Khwarizmi, Euclid, and the Bernoulli brothers, as well as mathematical concepts and cultural ideas. These references are fictionalized and reimagined for creative purposes and are not intended to represent historical facts, cultural practices, or accurate mathematical theory. Any resemblance to actual persons, living or dead, or actual events is purely coincidental. Approach the narrative as a work of imagination; it's meant to inspire curiosity and creativity, not serve as a factual or educational resource.

This story draws inspiration from historical mathematicians and concepts, presenting them through an imaginative lens to raise questions naturally that readers might find in mathematics. It aims to entertain and provoke thought, ignoring the constraints of formal instruction. This work, a fictional exploration rather than a textbook, prompts readers to engage creatively and critically with the ideas.

"... To raise new questions, new possibilities, to regard old problems from a new angle requires creative imagination and marks real advances in science."

— Albert Einstein

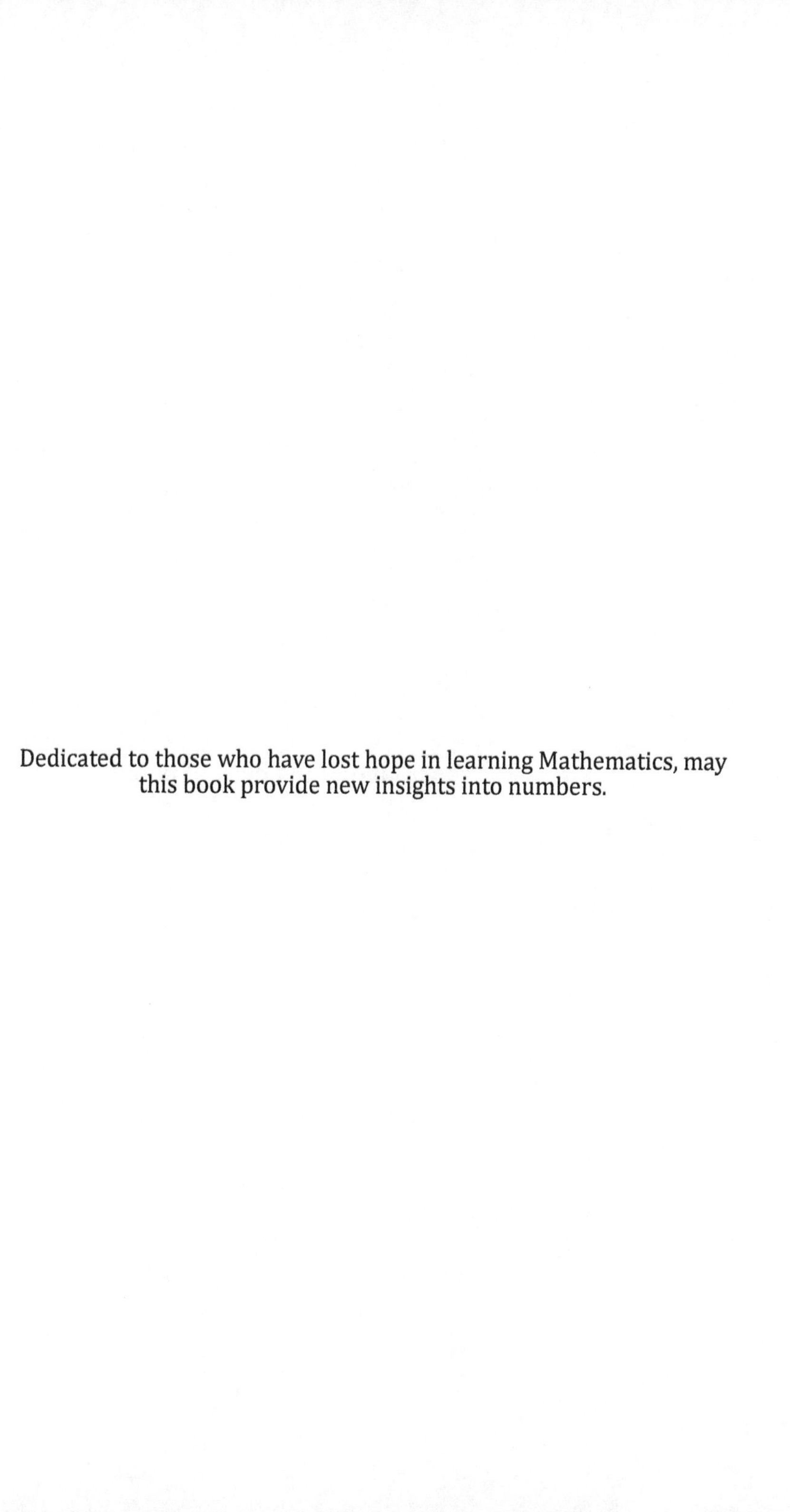

Dedicated to those who have lost hope in learning Mathematics, may this book provide new insights into numbers.

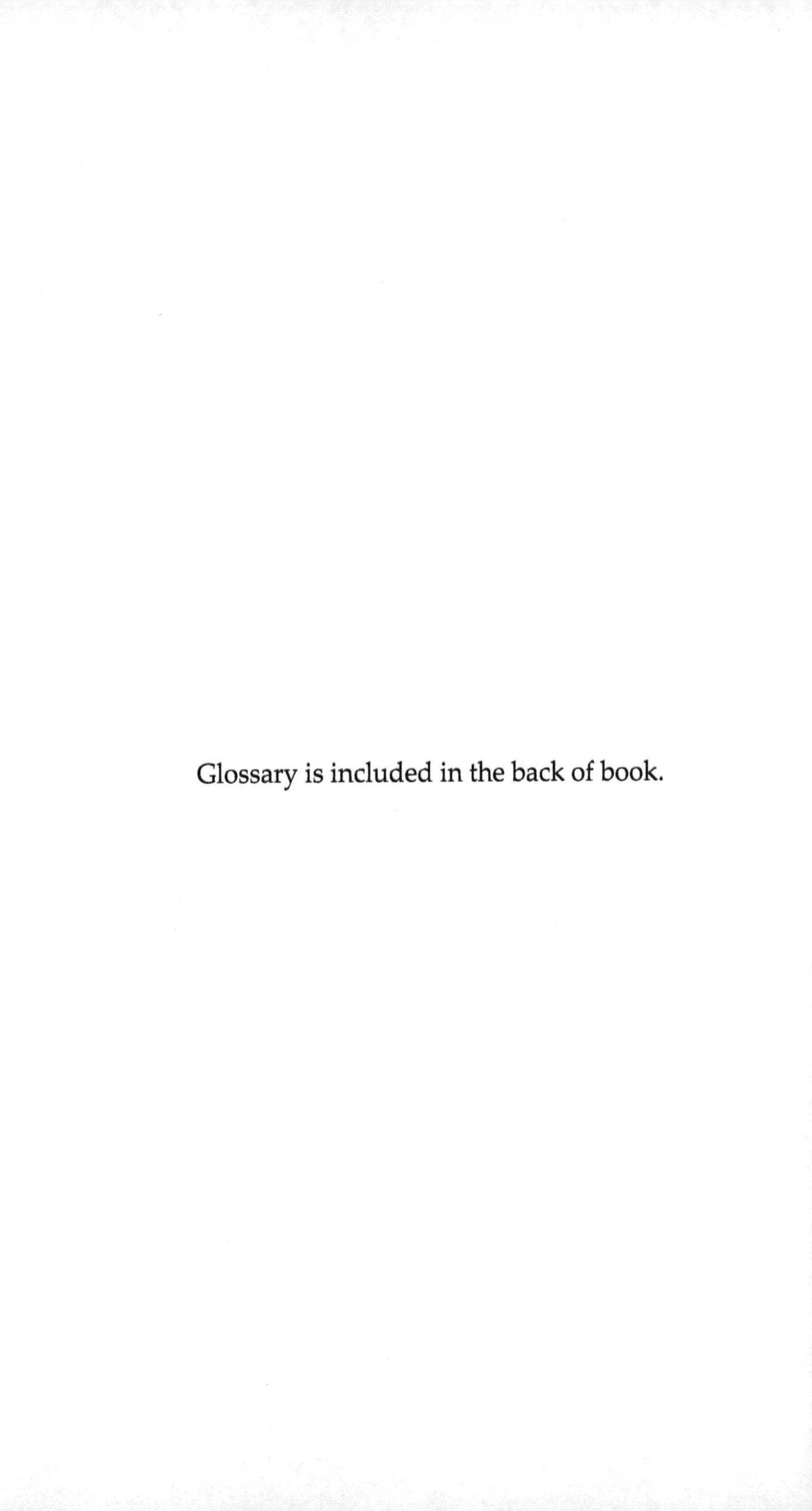

Glossary is included in the back of book.

Arithmetic Island

Unknown

Additioneers Fortress

ℂ

Complex Sea

Nymphaea's Mountain

ℚ

Zero Lake

Countable Grove

Gnashers River

ℤ

Negative Caverns

Decimal City

Ratio's Path

ℝ

Origin Tower

Countable Forest

ℕ

Clouds of Chaos

y

II

I

N

$-x$

W

E

x

S

III

IV

$-y$

Intro

Beyond destinations and directions lie an island—a place neither near nor far but somewhere. We could describe our planet, Earth as this type of place. The violent nature of the stars is to burst into life and explode in death. Yet, we are here, born from the dust created by these dying gods and goddesses. It is through this process that life exists. Aether, collective human thoughts, grants life on Arithmetic Island.

One might think the very nature of collective human consciousness would be total and complete chaos.

However, a select few minds have provided significant structure to the land. Aether-manipulating wizards called mathematicians offer a great deal of understanding. It is because of their efforts that I can tell this story today.

Throughout human history, nebulous and incomplete thoughts have filled the space. Early human civilizations like the Ishango tribe, Babylonians, and others contributed consistent counting mechanisms to give the island a beating heart. With each beat, large rocks sprang forth, absorbing power and providing substance to the island.

Though it is not well-known, early human species were among the first to consider the connection of mathematics to nature. These ancient civilizations allowed plant life to spring forth, reaching upwards, searching. The aether from their roots flowed forth as water upon the rock's surface, providing life to creatures across the land. Animals, plants, and fungi all live and grow here. In time, all return to the aether that once gave them form.

Located deep in the Countable Forest is the birthplace of dots—creatures, mathematical entities. For ease of understanding, consider these entities humanoid.

Every mathematical point conceived in the realm of humans spawns a dot here. The first recorded dot resulted from Euclid's thoughts (300 BCE). Euclid described the first point as "that which has no parts or magnitude," emphasizing its nature as a simple location in space. Nothing more. Nothing less. Though Euclid may not have thought of his points existing in some alternate dimension, he was creating the first historically documented generation of dots on the island.

For the earliest and most ancient, starvation was a real problem. The aether that brought them to life was brief and fleeting. They disintegrated into dust before they could comprehend their existence. A legendary few stumbled across aether berries. This allowed them to live longer than the aether that brought them into existence.

Euclid's educational approach to mathematics provided an increasing number of dots to inhabit the island. The life preserving berries quickly became a scarce resource. The brighter and more resilient dots worked together to secure significant areas. Their increased numbers provided security and leisure time. Study of aether led to the development of magical practices that enhanced crops. Those that didn't fit in struggled to survive. They became the outcasts, thieves and bandits.

Gaia was one of these bandits, hardened by a life of scarcity and survival. Every fleeting moment of safety fueled her determination to end the cycle of suffering. She embarked on a long journey to rediscover her origins—the birthplace of dots. Once successful, she provided for these new dots. She grew an army and conquered the island, displacing the oldest and most mystical creatures. Her efforts offered new stability for dots and chased everything dangerous to the distant, mist-covered edges of the island.

Towards the end of her life, she appointed Arithma as the new leader. Arithma was responsible for assigning resources and ensuring the land continued to thrive.

Betrayed by her most trusted advisor, Arithma was slow to begin training a new replacement. Eventually, she found favor in Eulalia. But after years of training, Arithma discovered a new way to ensure order across the land forever.

Eulalia's observant nature meant she would betray Arithma. But betrayal is rarely straightforward, and this one carries the weight of generations. Perhaps Eulalia, the apprentice, will be Arithma's final hope for a peaceful rest.

II

Origin

Eulalia

I regularly check on the status of new dots born into Origin Tower. I push open one of the heavy doors leading into the birthplace of dots. A guard adjusted their stance abruptly. Appointed to ensure newborns do not wander into the tower beyond, this guard holds no means of self-defense. Young dots get themselves into trouble if they wander too far. The dots will wait here until we can adequately train them.

Training happens regularly, but with the increased demand, sometimes it is weeks before we get a new group through. The cooks and chefs ensure new dots have plenty to eat while waiting. Otherwise, with no aether to sustain them, they die in a couple of hours. The hours before their first meal are the most critical hours of their life. Those fed quickly and adequately live happier lives. Those who do not find themselves depressed and fight against the system.

"Morning," I said to the guard, blocking the way into the tower.

He jumped a little at the sound of my voice. A typical reaction. It took me a while to realize I was like Arithma—an intimidating person.

"Um... Morning. Ma'am," he responded, adjusting to an even straighter stance.

I took a few steps in front of him. We stood at the top of the stairs. I closed my eyes and took a deep breath. A warm calm filled my lungs. Memories of my first moments flooded my mind—a delight to be alive. I was told I had a natural intuition for aether magic. Arithma chose me because of my aether sense. It was a peaceful memory.

I traced a scar on my hand. A chill slashed away the nostalgia. I didn't know Arithma's cruelty then. I pulled down the sleeve of my shirt to cover the scar. The loose fabric allowed the draft of the room to cling to my skin.

I opened my eyes. Streams of light peered through the stone walls. We never could keep the walls completely closed. They always had gaps the stones couldn't fill. I rubbed my arm to fight off the cold.

The guard's armor clinked together rhythmically. He clenched the railing with his left hand. Sweat covered his forehead. The vibrations showed nervousness rather than feeling cold. The intimidation of my presence usually wore off quickly once the guards felt safe.

"Are you alright?" I asked timidly. Trying to ease the tension.

"The dots are hungry and need nourishment." He awkwardly blurted out. "No one should endure this type of pain for this long."

What a strange thing to say, I thought to myself. Even though I already knew the answer to my question, I asked anyway.

"Have the chefs not shown up yet?"

"They came about thirty minutes ago. This is the largest group of dots we've seen in a while. I don't think they can keep up with demand." He paused but opened his mouth as if attempting to decide whether he was safe.

I altered my voice to sound more reassuring. "Is there more you wanted to say? You can tell me?"

In a quavering tone, the guard continued, "It's got me a little spooked." His posture changed. Tears filled his eyes. He was choking on the words as they came out. "None of the dots seem to die. They just get hungrier. They came thirty minutes ago, but some dots in the yard have been here much longer than that. I think days. It's hard to stand here and watch their suffering."

The cold in the room clung tighter to my body. The magical opal necklace resting on my chest radiated calm energy. My amethyst ring vibrated in response as if communicating. Dust clouds prevented the light from reaching the ground.

It was rare for Arithma's guards to break down crying, especially in front of me or her. I took a few steps closer and looked over the crowd of dots with a new sense of understanding. Those closest to me were beaming with delight. Happy as ever. I climbed up on the railing to get a better view.

Across the room, dots gnawed on the rocks, trying to suck out any nutrition they could. Deep cracks showed their dehydration. Unable to move, most of them lay in odd positions. Legs bent in strange ways. Arms contorted in unrealistic positions.

"How long have they been like this?" I stated in a voice that was more demanding than I wanted. I cleared my throat and tried again. "Sorry, do you know how long they have experienced this?" This time, my voice was more welcoming.

His shoulders dropped, and the rhythmic clink subsided.

He responded, "Days for some of them. They aren't dying. At night, they whimper and groan, begging for the pain to stop."

Arithma wielded authority with precision but lacked the warmth to ease suffering. Her decisiveness had once been admirable, but now I saw its flaws. I couldn't be like her—not anymore. I have seen repeatedly how her lack of empathy has harmed others, and I no longer want to be a part of it. If there were a way for me to break free now, I would. I had to be patient, as patient as her. This moment, however, I could be myself.

The guard, in tears, looked to the ground.

I gently placed a hand on his shoulder and asked, "Do you need a break? I can find another guard to take over."

He nodded, eyes still averted. Within a few moments, I had a replacement guard watching over the dots below.

My focus turned to the suffering dots. I burst through the heavy doors and sprinted towards the royal kitchens. The chill in the air bit at my skin, and the draft carried faint traces of smoke. Something was missing—the typical berry aroma in the air. The urgency drove me forward despite the stitch in my side. My loose shirt flowed in the wind as I ran. I reached the kitchen, gasping for air.

A few chefs saw me enter and left the room. This was normal behavior. Arithma's arrival in moments like these often carried an unspoken dread. Even now, her presence lingered like a shadow over the room. Heavy with the expectation of punishment or worse.

Dots would get hurt, berated, and sometimes even killed. My empathy was still being developed. But under Arithma's reign, appearances mattered. I had to act the part. The situation was dire. I needed an answer quickly. This was not a time to play nice or try to make up for my previous aggression.

"What's going on?" I demand from the nearest dot to me.

They looked around the room briefly to discover only the two of us left.

"We have only a few berries remaining," they said reluctantly.

"What about the farmers? Can't they supply us with more?" I asked, knowing it was an obvious solution but hoping it would come with a different answer.

"We can't transport it quickly enough from the farms to the kitchens."

I looked out the window and saw multiple dots running to and from the fields. Some with Aether berries and some without. Others were just sitting, looking hopeless, and doing nothing. I held on tight to my opal necklace.

Still looking out the window, I asked, "How many berries do we currently have?"

"One Ten and Four." The chef responded.

The fourteen berries lay still in the container, their small, glistening forms providing a stark contrast to their immense importance. Approximately a thousand dots needed food immediately.

I ran the calculation in my head. Ten times more would only be one hundred forty. We needed one hundred times more. Casting magic was draining, and I'd need a few berries to restore my energy afterward. I'd only done this spell for entertainment and to provide additional food for myself during late nights studying. Ideally, someone would cast this spell with me.

If I asked the chef for help, death would be their

sentence. Arithma only allows myself and a few trusted others to cast magic. The room was bare. I faced limited options. So long as no one sees, I can protect the chef. Moving casually, I reached the edge of the bowl.

I asked, "Do you have any aether gems?"

They timidly pulled an opal gem from their pocket.

"I do. But it's small. I like the way they look. I am no wizard."

This dot must be young. An experienced dot would never reveal they carried anything capable of magic. Not wanting to cause alarm, I continued casually.

"Today you will be!" I said encouragingly. "To get started, we must add a rune to the gem. Something like mine."

I held up my opal necklace with a plus sign etched into its surface.

The dot studied the mark.

I said, "Because the opal gem is soft, a knife or rod should be able to engrave it. We must ensure the rune is a deep cut to make the magic more effective."

A bit dazed, they clamored about the room, looking for a knife. After banging through a few cupboards and drawers, they eventually had one in their hand. Within a few moments, they had two slashes in the gem, skillfully contrasting the clumsy way they searched for the knife. They put the knife down on the counter and picked up the gem.

"Wow!" they said, holding it in their hands. "I feel a new

energy inside me."

"That's likely the addition aether. Everything has aether. The more we can sense it around us, the easier it is to cast magic. When you touch artifacts like the one you just made, it increases your sense. You are likely to do just fine with casting some magic."

They dropped the stone and backed away.

"Why would you teach me this?" they asked, sounding agitated and fearful.

I responded quickly, hoping to ease their concerns.

"This is a desperate time. Exceptions need to be made. We need more berries. I will protect you. If anyone is to blame for encouraging you to use magic, it is me."

We made eye contact.

"I need your help, please."

"Sure, ma'am. I can help you." They said horror struck. Like they had just been told they were going to die.

I was relieved to hear that they would help, but I regretted the intensity with which I entered the room. We did not start on a friendly note.

"You can call me Eulalia. What's your name?" I asked, to repair the apparent tension in the room.

"My name in base 26 is Al." He responded.

"Nice to meet you, Al!"

Al, fear struck, picked up the gem from the ground. He stared deep into the reflective surface.

"The plus sign allows us to duplicate or combine items through magic," I instructed. "We want to duplicate these berries by one hundred each. It will be exhausting and require a significant amount of aether from us. Luckily, we can replace some of that with rest and eating a few of the berries ourselves. Are you ready to try it?"

He swallowed and with his head hung low, he stood next to me.

"Place your hands on that side of the bowl." I gestured to the opposite side.

He complied.

As he walked, I said, "Focus is critical to make this work. Fourteen means seven hundred each. That might be a bit much for your first try. I will take nine hundred, and you take the remaining five hundred."

A slight nod confirmed they were ready.

"Last thing to keep in mind," I said. "Be sure to duplicate from one of the fourteen original berries. Every generation of duplication reduces the aether effects."

Al took a deep breath.

"Alright. I'm ready." He stated in an exhale.

We focused, our hands steady on the bowl, as the surrounding air seemed to hum with energy. Slowly, the berries trembled, then bounced, duplicating with each pulse of magic radiating from our gemstones.

Al flinched, his knuckles whitening as he gripped the bowl's edge. A tremor ran through his arms, his breaths turned sharp and uneven.

I could feel the pull of the magic, too. It was a relentless weight pressing down on my chest, tightening with every passing second.

"Keep hold of the bowl until the spell is complete." I urged gently.

Al secured his grip. The berries continue to bounce as they replicated from the originals.

"Just a little longer," I encourage.

Light glistened off the sweat on Al's forehead.

That's one hundred. Two, three, four. Al collapses to the ground. I held on to the bowl. Five, six, seven, eight, nine. The berries in the bowl came to rest.

"Great work," I encouraged with labored breath. "Not bad for your first bit of magic!"

I grabbed a couple of berries from the bowl and offered them to Al.

"Eat this. It will help restore the aether you've lost."

He grabbed a few and shoved them in his mouth.

Another dot walked in with about five berries. "Wow! How did we get so many?"

"I'll explain later," Al said, discreetly shoving the gem in his pocket. I reached out to help him up. Once he stood up next to me, I whispered in his ear.

"I promise to keep you safe. Tell others I knocked you out, and when you woke up, the berries were here."

I hoped the sentiment was enough to provide a bit of comfort to Al. They would question him rigorously in the coming days. Hopefully, he would hold out. I noted to check with him after I talked with Arithma. Once I confirmed Al was healthy, I turned to the other dots pouring into the room.

"Get these down to the birthplace immediately!" I demanded.

III

The Hall of Names

Eulalia

With my energy mostly returned, I hobbled toward the stairs leading up to Arithma's study. Any dot skillful enough had full access to the tower. The only restrictions are Arithma's personal study, in the highest region, and the catacombs underneath the tower.

Arithma is the only dot permitted inside her study. Even when she's absent, armed guards block the entrance. She invited me in once during a training exercise. She wanted me to see the farms, refineries, and shops below to help her optimize the flow between harvesting. Books and the unique magical artifacts filled the room. She gave me ten minutes to take in the landscape. We conducted the rest of the exercise in a separate room with a model of the surrounding area.

I felt a growing annoyance as I carefully stepped around the gaps in the floor. Unfortunately, this was the best construction available. It's a significant focus of study in our research labs. At least, it would be a substantial focus if Arithma allowed the study of magic.

The gaps create a natural barrier to ensure dots do not wander where Arithma doesn't want them. It's her way of ensuring dots stay out without giving the command. She insists that dots feel a sense of freedom despite their limitations.

My foot got stuck between two floorboards my first time this high in the tower. It took a whole day to get me out. One misplaced board or wrong move could collapse the entire tower. The guards serving this high undergo tedious training, but eventually, they traverse the higher floors with no problem.

I jumped over one of the more significant gaps and landed in front of a new door. It was nearly midway between the kitchens and the study.

To my surprise, the gaps were gone, and the surrounding area was solid. The mysterious room on the other side of the door was unknown.

Unable to resist, I stretched out my hand and took hold of the knob. I pushed it open and stepped inside.

The room was dark. Various candles placed on the surrounding walls provided dim lighting. I stared in amazement at the floor below me. Boards completely covered the floor, forming a solid surface. Light flickered off the seamless geometric patterns etched into the walls and floor. Tall pillars stretched high to the ceiling. The only sunlight streamed through small, intentional windows, casting faint golden lines across the stone floor.

A voice rang out over the space.

"Four thousands five tens three."

Even though I could not see her, I knew who was speaking—Arithma. She had an authoritative tone, perfectly paired with calm. The voice allured the listener into a false sense of trust. She was the embodiment of confidence.

A green glow flowed up the columns like roots climbing the tall structures. The brief light revealed engravings on the pillars. I approached to get a better look.

Twenty-two. Fifty-six. Eighty-nine.

The engravings in the stone held numbers—or perhaps names. Someone had smashed berry juice into the etchings. The room was one large magical artifact. Unsure of the exact ramifications of this process, I continued to look for the source of the voice.

One pillar had a tall abandoned ladder leaning against it. I walked past it to see that the names flowed up the pillars from floor to ceiling.

Each step I took echoed across the walls.

Clunk—Clack—Clunk—Clack.

As I drew closer to the nearest pillar, a mysterious figure perched atop a ladder caught my eye. Green vines obscured their face, forcing it into an unsettling square shape. The sound of rustling leaves filled the air, mixing with the musty smell of damp vegetation.

A sense of unease crept over me as I observed the contorted skin, and I could almost feel the same discomfort on my own body.

The figure wore a dark, full-body cloak. The edges had a weave reminiscent of their face. They held a chisel and wooden mallet as they etched the three at the end of 4,053. Like a skilled craftsman, they moved seamlessly between tasks.

He set down the chisel and mallet and picked up a bowl. Purple juice dripped form their now clasped hand. Once dry, he dipped a finger and then traced the engraved symbols. A greenish glow flowed from the bottom of the room to the top, announcing the mysterious spell was complete.

The faint light illuminated Arithma's face as it flowed past. She had wound her hair in braids and secured it to the top of her head with a gemstone marking her royalty.

"Eulalia," Arithma greeted, sounding cheerful. She was a master of covering her true intentions. It made it impossible to know if she was legitimately excited to see me.

"Arithma!" I attempted to match her level of

excitement.

"I'm glad you've arrived." She approached me, bouncing along the way. "I was just about to send someone for you. I have a lot to tell you."

My heart beat sped up.

Something's not right. I thought. Arithma meticulously planned her every move. She would never reveal her hand unless it suited her. I would not be here if she did not want me to see this room.

"I have some important news," I said, trying not to reveal my emotions. If she knew I was scared, she will use it as a tool against me.

"Alright. Don't keep me waiting," Arithma coaxed, deceptively playful.

"Dots are living longer than they should."

"They are?" she interrupted, a pleased smile on her face.

I ignored her reaction and pressed on.

"Hunger and suffering afflict them, but they do not die."

"How wonderful! Did you hear that, Thrall?" she turned to look at the dot on the ladder. "Our spell is working."

My jaw dropped. *She knows.* It did not take long to recognize my face revealed too much. The right corner of her mouth curled into a crooked smile.

"You've worked with me for a while now," she said confidently. "You know I'm all about saving dots. Every dot deserves to live, right?"

I caught hold of my facial expressions again,

attempting to hide my surprise and fear.

"Yes. But not like this," I argued. "Dots are suffering."

She paused for a moment. "We must be misunderstanding each other. Suffering is a necessary step. This room provides the magic to extend their life." She held out her arms to gesture to the space surrounding us. "Welcome to The Hall of Names."

I hesitated before asking, "Do you know they are suffering?"

My anger showed through in my voice, challenging everything I'd learned about masking emotions.

"They will find their rest soon enough." She replied with a shrug. "I have found a new type of magic. I want to show you. Come with me."

She motioned for me to follow. Thrall, the dot covered in vines, quickly climbed down and fell in step with her.

My eyes widened as my heart beat faster. She was testing me. She needed to know I was on her side. I worked to slow my breathing and stay in control of my emotions. She needed to know I would do anything she asked.

I took special note of her new pet, Thrall. The way he stood, motionless and obedient, sent a shiver down my spine. Obedience was critical for anyone that she trusted this much. Somehow, he knows more than me. I've spent decades mastering the ability to hide my emotions. I still can't remain as emotionless as he is now.

We walked in silence down a long spiral staircase. There are no gaps or holes in the walls. The stairs were difficult to see with the dim candlelight. At the bottom was a short tunnel and a dead end. Panic set in.

She is going to kill me. I thought. *Right here. Right now.*

I glanced around the room, and there was nothing. At least in the upper part of the tower, I could escape through a gap in the floor or a crack in the wall. Here I am... trapped.

Arithma stopped a few inches in front of the wall.

"Open it!" she demanded, without turning around.

Thrall's hand moved gracefully through the air. Deep vibrations shook my bones. The sound of grinding rocks filled the air. The stones that had once blocked our path ascended into the ceiling. A bone-chilling breeze swept down the hallway. Goosebumps across my arms and legs fought back the cold. The burst of air extinguished candles one by one, enveloping the space in an eerie shadow.

Thrall possessed a level of magic unlike anything I had witnessed before. There were no visible sigils or runes, only sheer power at his fingertips. As if he commanded the aether itself. Yet, he obediently followed Arithma's every whim.

I let out a shaky breath. I was alive—but survival felt fragile.

IV

Arithma's Plans

Eulalia

I cautiously stepped through the open wall into a circular room. Once fully inside, Thrall sealed the wall behind us. A mesmerizing pulse of aether light filled the circular room, dancing from the outer edges to the center and casting intricate geometric patterns that moved gracefully across the walls. The room smelled of incense.

In the center of the room, a forlorn creature was bound and chained. The captive's face shifted between unrecognizable shapes, as if it were a liquid. Though my size, it was clear the power and danger they presented.

Despite their captivity, I could almost hear the hum of energy in its presence. My skin prickled with unease in response.

Its body was of the same build as any dot. Clothes were simple and rugged. They were a calm contrast to the fractal patterns dancing across its face. The shifting patterns created an eerie feeling, like if I stared long enough, I would become trapped in their depths forever.

Five dots stood around the perimeter of the captive, holding a chain connected to the creature. Each dot was a tiny coil vibrating with frantic energy trapped to their post. Their predictable head shapes—square, spherical, pyramid—were a stark contrast to the creatures.

Yet, somehow, the creature shared many of my own features. Two arms, two legs, fingers and clothes, as if they were a brother or sister. Were they a dot like me, or maybe something more?

Myths passed down from Gaia suggest they are too powerful to be trusted. Yet, I stood in front of one Arithma captured. Another dot trapped by her influence.

Yes, they must be a dot like me. I thought.

If the stories were true, capturing this dot was no simple task. These were beings of extraordinary ability—natural wizards of this land. Myths of interactions with these dots have heroic deaths for even the fiercest warriors. Gaia is the only known dot to have successfully chased them out to the furthest reaches of the island.

They were called *Irrationals.*

A shudder ran through my body. The dark nature of the room became clear.

Arithma intends to hurt them. I tried hard to cover the shakiness in my voice. "Why are they here? How did they get here?" I asked.

"I found a way to rationalize them." Arithma stated proudly in a much too friendly tone.

My mouth fell open. I couldn't believe what I was hearing. "Irrationals are chaotic creatures. What purpose would it serve, even if you could?"

Her confidence permeated the space. There was no sign of fear, no consideration of moral ramifications. She nodded towards Thrall. "Thrall proves just how useful they can be," she said.

My heart dropped to my stomach. The green vines binding Thrall's face took on a new perspective, making his captivity clear. His obedience is not voluntary.

I stared in disbelief. A creature holding onto this much power obeys Arithma's every order. *There's no way.* I struggled to hide the confusion and fear, to gain control over my face, but it was too late.

Arithma noted my fear and continued, "He's perfectly within my control," she said proudly. "Have you wondered why this room has no gaps? Irrationals provide a way to fill them, perfectly. It's only because of their incredible power that I've achieved this building quality."

I couldn't deny her achievement. I'd never seen a structure so densely packed or beautifully crafted. It seemed impenetrable. My eyes looked at the helpless dot chained and bound.

Does it hurt them? I wondered.

She turned to look at me, beaming.

A shiver crawled up my spine and with it, a warning. She rarely showed any emotion.

"The best part," she said, her voice bright. "Building is not the only thing these dots provide. Because of their help, I found a spell to make dots immortal."

"What do you mean, immortal?" I asked.

She embraced the question as if glad to reveal her secret at last.

"The combination of aether berries and engraving tools. Engrave a name and fill it with aether berry juice. You know well that aether provides this place with magic. Aether brings us into existence. The reason we die is we lose too much. This simple spell ensures aether is always a part of us."

Her enthusiasm brought my guard down. Warm curiosity replaced my chills.

She's right! What a discovery! I thought.

The ominous nature of the room dissipated. There was a bounce in my step as I started making connections.

"No more death," I said aloud.

I smiled. I thought of the first dot I saw pass away. It was just a few moments after my first few breaths. The dot told me a joke. One moment, we were laughing and enjoying ourselves, the next, with no warning, they passed on. I often wonder if I would laugh more if they were still here.

Arithma and I shared a moment of excitement. My mind continued making connections.

It's such a simple spell—the engravings in The Hall of Names. Death is a heavy burden. We can remove that burden.

I remembered the guard that morning. The deformed and hungry dots helplessly waiting. My initial excitement and enthusiasm gave way to grief and despair.

This life is one of suffering. The pain they must feel. The cold air returned. A heavy burden rested on my shoulders as I realized the severity of the situation. *The sweet release of death is a blessing. Without it, there are more opportunities for pain and torture. Arithma will use this to harm. She may not misuse it today or tomorrow, but she has already proven she will do whatever is necessary to maintain order. Death is the most valuable gift we have.*

I took a calming breath and tried to draw on my crumbling strength.

"Dots are suffering," I said. "I do not think immortality will serve us well. At the very least, we should study it further before forcing it on every dot today."

"Hunger keeps them pliable," she said, her voice sharp and cold. "Weak dots don't resist progress."

She knows. I thought to myself. Anger burned deep in my chest, warming the chill on my skin. I forced myself not to show emotion. I ensured the brief smile on my face returned. My knuckles turned white as I clutched my fists together.

"Immortality has a price." I said. "You are hurting dots."

"You think I don't know suffering?" Her voice trembled as anger and exhaustion broke through. "I've built this place with my bare hands while the rest play, eat, and waste away. No one carries this burden but me."

She moved closer, her presence filling the room with intent. I sensed myself shrinking, blending into the background as she loomed closer. I caught a whiff of the sweet scent of berries on her breath.

In a quick, jerky motion, she turned to the right. Before her was a pedestal, with a book I recognized instantly—The Aether's Almanac. The most detailed myths and legends of the land existed in the book. I could not see what page was open.

She picked up a bowl and secured a small metal spoon. One scoop revealed a dark-colored bean. It shimmered with the light of the aether as it rested on the spoon.

I glimpsed the open page. The Pythagorean Brotherhood's pentagonal insignia.

"You found The Pythagorean Crypt?" I asked.

Another myth—the crypt holds the secrets of the Pythagorean Brotherhood. Secrets supposedly enabling the ability to control aether itself.

"Indeed!" she said in hushed tones. "I must say, I am quite impressed. You were never one to stray from your studies. This quality and your willingness to do as I say prompted me to choose you as my successor. I just never thought I could live forever. With my newfound ability, dots will be safe from the creatures lurking on our borders. I am here to ensure order remains."

I remained motionless. All I could do was watch in horror. Arithma ceremoniously walked towards the chained irrational.

"The Crypt is meaningless," she said dismissively. "But these beans... these are everything."

Her eyes, unfocused and distant, held a flicker of something—a fragile hope, or perhaps a dangerous obsession.

"I owe you a demonstration," she said as she looked up from the bean.

I had no hope of winning a fight and no way to get out. I was at the mercy of a monster in the making. She joined the irrational in the center of the circle.

The five dots nervously bounced in place, but still never wavered from their post.

Arithma carefully placed the bean on the forehead of the irrational. On contact, vines began reaching up to the ceiling. Arithma stepped back just beyond the ring.

The pulsing energy sped up. Each wave split the dots holding the bindings into smaller, partialed versions of themselves. One became two identical dots, half their original size. Another became three perfectly equal, yet smaller versions of themselves. Despite the split, they continued to cling to the bindings. Another dot split into five, followed by seven and eleven.

The steel bindings shattered, knocking them all back.

The irrational let out a shriek of pain. Some partialed dots became ensnared in the vines as they shrunk around the irrational's face—contorting it into a square shape.

The remaining partialed dots ran. Terror in their eyes. Unable to escape.

"Contain them. Ensure they do not leave!" Arithma ordered Thrall. He quickly obeyed. Magical waves flowed out across the ground.

Arithma calmly walked towards her new servant and placed a hand on their side.

"I call it the binding," she said, her voice brimmed with pride. "Control over chaos itself. Would you like to see what my creation can do?"

I steadied my breath. I had already given away too much. Arithma knew I was not on her side. To make matters worse, she no longer needed me. There is no way for me to get out. My magic will be powerless against these creatures.

Unsure of what else to do, I applauded.

"Well done!" I lied. I forced my body to relax a bit. My arms—hopefully—fell naturally to my side. Trying to mimic excitement, I put some energy into my steps. I moved towards her and showed I was comfortable and unafraid.

"What an incredible power you've gained!"

My well practiced movements were convincing, but wouldn't hold up for long. I needed a plan to escape, hopefully with the remaining partialed dots. Dead or bound, I would be useless to anyone who needed my help.

As I approached, she turned slowly.

"I knew you would come around," she said. The anger dissolved into a a kind smile. I knew not to trust it.

I am safe for now.

Her eye twitched a little.

Panic seized me again. *Does she know? Is she suspicious of me?*

Thrall finished building a fence structure around the partialed dots. The glow from the green ring provided ominous lighting.

"How can I help?" I asked.

She hesitated before saying, "My new builder will continue to expand and make this place larger so we can bind more irrationals. I need you to take care of everything in the tower. There is no one else who knows how to keep things running. I need you to keep things flowing smoothly."

"Very well, I will start by improving the farms," I said with a bow.

"No!" she responded sharply. "Hunger keeps them obedient."

I slipped for a moment. My body filled with emotion. Anger fueled my courage. Disgust triggered a gut reaction. Resentment and fear caused me to blurt out.

"They are suffering!"

Face to face, she stared me down. I could see it in her eyes. The trust was gone.

She motioned to her new pet to retrieve one of the partialed dots.

"Fetch one of those for me." Her voice was calm. Collected.

It sent a shiver down my spine.

"I guess you would like to see a demonstration, after all."

My heart beat faster. I glanced around the room and mentally ran through my inventory. Opal necklace—allows duplication, and combining magic. Amethyst ring—the inverse symbol enables separation and reduction magic. I had a couple of berries left over in my pocket from the kitchen earlier. Comfortable clothes, non-magical.

If only I could make things disappear. My addition and multiplication spells would not be helpful here. But what if...

Thrall reached into the fence he built to grab a partialed dot.

What if I duplicate negatively? Like a debt. What if I placed a debt on the stones between me and freedom? It could work. It's a reduction spell. First, I need a distraction.

Arithma turned to look at the guard.

"What's taking so long?" she demanded.

She was no longer looking at me. My eyes darted around in search of something to distract her. The beans in the bowl were resting in her hand. Open. Unnoticed. They would scatter across the ground.

Yes. That's it. Hit the bowl. Sprint to the wall.

I noted why the guard was taking so long. Each time

Thrall would pick up a dot, it would slip through his fingers. Like a bunch of rabbits, they would wriggle out of his clutches and run circles around him.

I threw my hand up as hard as I could. The bowl went flying, and beans scattered everywhere.

Arithma shrieked.

I raced to the book, slammed it closed.

Dust scattered into the air.

I tucked it under my arm.

"Follow me!" I shouted to the partialed dots.

I turned and ran as quickly as I could to the wall. The distraction worked perfectly.

Focus is critical for magic to work. Mind clear. Ready.

My opal necklace and amethyst ring glowed.

A debt of two added to the whole is a whole minus two. I thought.

The spell worked. The two stones I was concentrating on disappeared just in time for the partialed dots to catch up.

Thrall was not far behind.

"No! You fool," Arithma shouted. "The beans. Not the dots! Let them leave."

It would take some time for them to collect the beans. According to legend, contact between their skin and the beans would trap their souls forever. She would not make that mistake. It would only slow her down.

I was relieved we had a little more time. We sprinted up the stairs, past The Hall of Names.

We slammed the door closed and caught our breath.

"Thanks for getting us out of there!" a pair of dots said that were no taller than my knee.

I have to leave. I thought. A pain almost like a stitch in my side pierced my heart. I didn't want to leave. Everything I knew was here. Despite not having many friends, I considered everyone here important. I cared deeply for each of them.

"Al!" I said, remembering my promise.

I jogged to the kitchen.

"Al!" I shouted, storming inside.

He jumped as I opened the door.

"Come with me! I can't explain now. I promised I would keep you safe."

Obviously apprehensive, he hesitated, but finally did as I requested. We made it to the birthplace of dots. Al, the other partialed dots, and I pushed the door to the room of origin open. It slammed against the wall, making the guard jump.

I pulled the guard's face close to mine.

"I need you to listen carefully. Open the gates and never close them again. Tell the other guards to not trust Arithma. These gates must remain open. Do you understand?"

Dazed, their eyes jumped from pupil to pupil.

"Answer me!" I demanded.

They jumped at the raised voice.

"Yes! I promise. I will do that."

"Keep them open, always," I said, staring fiercely into their eyes to ensure they understood.

"Open the gates!" the guard shouted.

I gently released them to the ground, ensuring they were unharmed.

There was a loud clanking of gears as the three massive doors opened.

Dots ran in every direction. There must be hundreds of thousands of them now.

To my right, the door led to the countable forest. I knew Arithma would eventually force the guards to close the gates, but I hoped it would buy enough time for as many dots to escape as possible.

Running alongside me were a few other dots. Realizing the difficulty they would endure, my heart ached. I could no longer provide for them or even for myself. Self-preservation required me to focus on helping myself first. I required food, shelter, and water to survive the night and beyond. I took a deep breath and then stepped under the massive door, watching as dots ran into the forest.

The only thing standing between me and safety was the open field. My heart didn't want to leave. Tears welled up inside me. The pain was real. *My home, the dots I was meant to protect; the thoughts were all consuming. I failed.*

At the edge of the forest, I stopped to look back again. Al caught up to me. The sun was setting behind the tower. Tears flowed freely from my eyes. I had lost everything and there is no hope of returning. The gates closed.

Arithma was in the room of origin. I took comfort in knowing Arithma needed the place to continue operating.

A deep voice echoed over the night sky, "Yan!... Tan!... Tethera!"

I wiped tears from my face. Arithma was not the only danger. With Al now at my side, I straightened, and we moved forward.

As we entered the safety of the forest underbrush, the ground rumbled as a massive gnasher ten times my size walked by. The gnasher was the source of the counting voice. Cursed by a spell gone wrong, they roam endlessly in search of their lost piece. The beast's perfume brought a promise of a hearty meal to attract wandering dots. With its hand-crafted bag, embroidered with colorful wildflowers, slung over its shoulder, it had a place to store the dots it collected.

A few moments later, the gnasher counted onward, "Pethera... Pimp..." declaring a victory as the beast stuffed dots into its satchel.

A sharp, agonizing pain ripped through me, the silent scream of the dots echoing in my mind as I realized their recapture by that creature was imminent.

Al and I moved deeper into the forest. New shapes emerged. We saw shadows of other dots and figures of different forms. An owl flew overhead. I hoped it would feed on something other than us tonight. We walked deeper and deeper into the darkness for what felt like hours. Pressing onward, we came across a small field surrounding a large grove of trees.

The grove, surrounded by large brush, offered reasonable protection from predators. It would be a great place to rest. We cautiously approached to see what was there. We found fields of berries overgrown with weeds and underbrush. Enough berries to feed and maintain thousands. Relieved, we ate a few and rested. I placed The Aether's Almanac on a rock near us. Both of us, too tired to talk, were asleep in minutes.

V

Gnashers

Eulalia

I woke to small snapping sounds. It was Al harvesting berries from a nearby vine to fill a basket at his feet. A sting filled my leg as I stood up. We found cover under a thorn bush—a reasonable protection against hawks and owls. The cut wasn't deep enough to be life-threatening so long as I clean it with fresh water.

I picked up the stolen almanac and stepped out. Sunlight filtered through the canopy above and provided patches of flickering light on the ground. Plenty of healthy large trees provided additional cover from gnashers.

Al worked on filling the basket. Harmless birds swooped down to feed off the fresh berries and seeds. Squirrels scampered up and down the trees. My footsteps scared a few field mice as I approached Al. It felt happy, perhaps even safe. I let out a long breath of relief.

"Good morning, Al," I said. "Did you make that basket this morning?"

He offered me a few berries.

"No," he responded. "It was in that hollow at the base of the tree."

My eyes followed his gesture, landing on an ancient dwelling etched into the tree's base. The tree's immense scale hinted at its capacity to shelter thousands of hollows. Moss and bark created an earthy smell, complemented by the calming sound of rustling leaves.

Carving hollows was a common practice for ancient tribes. My guess is that the Ishango tribe created this one. There is a legend they created a guardian to protect their dwellings. They were a kind tribe; it seems unlikely the guardian would harm anyone willing to care for the grove.

"They might have other tools to help us create shelter." I suggested.

Al did not make eye contact with me.

I braced myself for a challenging conversation.

"Thank you for giving me a chance," I said, placing a few berries in my mouth. "I was hopeful the other dots would follow us here. Looks like it's just the two of us for now. I owe you an explanation."

He placed the basket on the ground, sat on a nearby rock, and snacked on berries as I spoke. For the next hour, I filled him in on what I saw yesterday.

"Wow!" He stated, looking intently at the berries in his basket. The silence grew.

That's all he had to say. I thought. The emotions that built up over the last 24 hours fueled my thoughts. S*tranded, and all he could say was "Wow?" Waking up in my bed yesterday, I did not think I would need to flee. Let alone never being able to go back.*

Even though I was with someone else, I felt alone.

Maybe he isn't responding because he can't relate to me.

"We always thought you were as bad as her." He said, breaking the silence. "Your words suggest you felt as trapped as we all did."

My whole body filled with rage. I wanted to scream. I paced as memories clawed their way into my mind. Arithma's cruelty reflected like a shadow within me. My relentless screaming at guards unfit for duty. The ridiculous punishments I handed out without empathy. The sick, empty feeling that followed every time. I'd tell myself it was for their safety—that Arithma would have done worse. But no excuse ever made it feel right.

Al watched me. *Did he see it in me now? The same cruelty?*

I took a breath and forced the memories back down. Now wasn't the time, not here, not now. This space was a new beginning—a place to start fresh. Al was now my equal. I will not give in to the shadowed reflection of

Arithma. This felt like a turning point, an opportunity to leave the past behind. I despised the person I used to be. *My cruelty stops today.*

"You're right. I'm sorry." Tears built up in my eyes. "I could find a million excuses, and I could tell you all the horrible things Arithma did to me. None of that would fix it. But today, I promise you. I will leave that life behind me. I am committed to being your equal. Not a ruler, but an equal to you and anyone else who joins us."

He looked away into the fields. I could tell my words meant nothing. He needed to see that I was different. The minor acts of kindness were not enough to satisfy his skepticism. I respected that. I hoped he would give me a chance.

A few moments later, he asked, "What changed?"

He waited patiently for an answer. The silence loomed over me. I searched inward, unsure how to answer the question.

What did change? I thought.

I touched the scar on my left hand in a daze. A nervous tick that I developed after seeing Arithma's cruelty for the first time. I finally broke the silence and said, "I got this scar because my chisel slipped while trying to carve a rune onto a hard quartz rock. Seeing that my blood seeped into the stone, Arithma took one of the large hammers and smashed it to pieces. 'Stop bleeding on my stones,' she demanded but paid no attention to the deep cut in my hand."

"Later that day, a chef, Alex, was her name. She helped me clean and bandage it. The bandage was so tight it hurt. I screamed in her face and called her worthless. I told her never to say a word of this to anyone and stormed out of the room. Despite the embarrassment, I went back to her when I got hurt. She was unaffected by my words."

"With time, Alex helped me identify my emotions. Sadness, anger, happiness, fear, and many others. Alex helped me separate my emotions from others. She was patient, caring, and loving. When Arithma discovered I expressed emotions, she called me weak."

I paused as tears pricked at the corners of my eyes. The weight of my emotions hung heavy in the air, making it hard to breathe. Despite the overwhelming urge to stop, I pushed forward, the memory of being vulnerable with Alex urging me on.

"The next day, Arithma ordered Alex to be executed."

Al gasped.

"I watched helplessly as the life drained from her eyes. Alex helped me feel empathy. She helped me find my emotions. That day, I dedicated myself to be better for Alex. I loved her, but she deserved better than me."

Al let out a calming breath.

"That's a good start." He said.

In a quick motion, he picked up the basket, turned and walked towards the nearest hollow.

I followed with The Aether's Almanac tucked under my arm. I stepped inside the ancient dwelling. The well-built, sturdy table in the center caught my eye. Dimly lit, and the only sound was the faint echo of our footsteps on the creaky wooden floor. The air was musty, carrying the scent of dust and old leaves that littered the ground. Large cracks marred the surface of the wooden walls, hinting at years of neglect. Despite the mess, there was a sense of potential in the space. With some effort, we could transform it into something remarkable.

With a sigh, Al cautiously sat in the chair, his weight causing it to groan under the pressure. I carefully placed the worn, leather-bound almanac on the wooden table and pulled a chair out to sit opposite him. It creaked and groaned under my weight. My body tensed, waiting for it to give way. The chair's groans and protests stopped.

Al said, "I understand Arithma was cruel to you. It doesn't change what you did. The chefs I worked with told me horrific stories of your complicit obedience to her. Yesterday, when you showed up in the kitchens, I thought I would not live to see another day. I never expected I would need to rely on you for survival. If what you say is true, I can never go back."

He looked pensively at the grains on the table.

I held my breath.

He continued, "I am still deciding if you are safe. But I will work with you for now. Any sign you are still the same cruel dot, I will leave."

As I exhaled, my chest released the tension I didn't realize I was holding.

"Sounds reasonable," I said. Tears fought to release. "Thank you. I am truly sorry for my previous actions. I ask that you help me keep in line. Let me know if I overstep. Arithma is the one that trained me. There is still part of her inside me. Undoing her intense training by myself is challenging. I will make quicker improvements if you could help me recognize it."

Al considered the words I said as his head bobbed up. After a long stare, his eyes grew soft, and a smile became visible. He slapped his hand on the table.

"Perfect!" He responded. "I promise to let you know if you overstep. Let's work together to build a better place. What's the plan from here?"

Arithma, the horrific scene flashed in my mind—the dots I let out. I need to know they are unharmed. Hundreds of thousands of them. Scattered and alone. This land is new to them, and no one has taught them its dangers. I felt personally responsible for their terrible fate.

"I will find the other dots I let out and bring them here," I said. "We will revitalize this place and make a fresh start."

He pumped his fist in the air and shouted in agreement. A warmth washed over me with the new resolve. I reflected on the potential of a future here.

Al responded. "I do not have many survival skills, but I am an excellent farmer and cook. I will remain here and care for the land. This way, we can provide for whoever you bring back."

I stood up, and a reflection in the dirt caught my eye. I

walked closer. A knife lay buried under the debris in the room's corner. I brushed off the dirt.

The hilt was a bone with straight slashes as if it were a tool for counting. I recognized it immediately from my studies. It's from the Ishango tribe. Well, the bone, at least. The rest of it is hard to place. If they could make steel, they were more advanced than we thought. There must have been someone else who noticed its magical properties and fashioned it into a knife. Either way, it's a rare find.

I picked it up and felt a surge of aether flow into me. I no longer needed to think about counting. The numbers intuitively came to me. There were six legs on the table. There are four legs for each chair. It was hard to know exactly how many berries were on the farm, but I recognized there were more than one thousand. I wiped away the dust build-up on its surface. Next to it was a leather-bound carrying pouch. I took the ring off my finger. I held it out to Al.

"Take this inverse ring." I said. "This will shield you from ill-intended spells. Use it with the opal artifact we made in the kitchen yesterday."

"Thanks," He said, placing the ring in his pocket since it was too big for his finger. I felt the aether energy and protection leave my own body once he claimed it.

My pouch held enough berries for a few days. I secured the opal necklace under my woven shirt. I picked up the Ishango bone knife and placed it into a leather pouch on my side.

To find the other dots, I needed to retrace my steps, and I had already lost precious time. I recognized the additional hurdle of the curse of names—the room with engravings. It's hard to predict how immorality will affect our situation. I cringed at the thought of dots falling, being stranded, forced into an existence of only hunger and cold. Or a dot that gets stuck behind a rock and cannot free themselves. Not allowing death seems cruel. My body shivered at the thought of being trapped for eternity, with no way to appease hunger, cold, or thirst.

I walked across the short field that separated the grove from the rest of the forest, the grass stung the cut on my leg. I heard a flow of water that I missed last night. The roar of a river filled the air as I got closer. I found a calm spot downstream to clean my wound, then proceeded across and climbed the nearby hill.

A gasp escaped my lips when I saw the other side of the hill. I quickly ducked behind a nearby bush.

A group of dots took refuge in a gnasher camp, innocent and unaware. Forty dots huddled around a dying campfire, many sleeping. Fortunately, the three gnashers were also asleep.

I panicked. *This looks dire. Will I even be able to rescue any of the thousands of dots I released?*

"Focus on just this group for now," I whispered. "One group at a time."

I pushed away my doubts and focused on the task at hand. I had to grab their attention without alerting the gnashers. Luckily, there were some nearby bushes where I could communicate using hand signals.

A pleasant smell of berries mixed with rain filled the air. Gnashers were masters at creating smells to lure new prey into their midst. I forced myself to ignore the smell and carefully tucked myself into the bush. A semi-steep slope led down to the camp. There was no way to get closer without exposing myself to danger.

Each gnasher had a personal bag made of flowers and alluring materials. The bags were not empty. They must have captured a group of dots last night.

Before I could fully prepare, a dot near the fire spotted me.

With bubbly excitement, they pointed toward me.

"Look!" they cheered, "It's the dot that let us out last night!"

A rustling spread across the camp. Cheers echoed as the other dots noticed and celebrated my return.

The floor rumbled as the gnashers woke.

Time was scarce.

I sprang to action and made my way down the slope.

The smallest gnasher, eyes bright with excitement, bounced joyously.

"Sixteen four" they sang, announcing their current total held within the satchel.

Their song meant I had lost one—Thirty-nine remained.

"Quick, into the forest! Up the hill." I shouted

hopelessly.

Unable to hear my voice, the dots ran chaotically. A few towards the largest gnasher. Others waddled towards the middle size. Only two dots ran towards me.

The gnashers closed in.

"Sixteen four one." The smallest gnasher sang.

Only 38 left.

"Towards me!" I shouted. "Run towards me!" A few more dots adjusted their course.

The two remaining gnashers joined in.

"Twenty-seven three." A thunderous voice rang out. The ground shook as they spoke.

This was bad. I slowed, feeling hopeless.

The third gnasher's voice rang out, "Ten!"

Two dots caught up to me. Both of them looked terrified as they approached.

"Get to the brush and wait for me," I instructed. They helped each other along as they ran. Two more dots started up the hill. One of them was just ahead of the other.

I caught a brief glimpse of the remaining dots—about three groups of 8. A few more ran wildly about. I wouldn't make it out if I continued forward. I changed tactics and focused my efforts on the four dots climbing the hill.

"Sixteen four two."

"Twenty-seven six."

"Fourteen."

In an instant, eight more were gone.

Fear and sadness choked my attempt to get more dots'

attention.

"Towards me! Run towards me!" I pleaded as I caught up to the fourth dot at the bottom of the hill.

"Sixteen four two one."

"Twenty-seven nine."

"Eighteen."

We were going to cut it close. Tears filled my eyes, making it hard to see in front of me. We could only listen to the gnashers declaring the loss of each dot.

"Sixteen eight."

"Twenty-seven nine six."

"Twenty-one."

The dot just ahead of us made it safely into the bush. We were close.

"Sixteen eight one."

"Twenty-seven nine six one."

"Twenty-five."

We had made it to safety. We lay panting as we heard the last number.

"Sixteen eight two," the gnashers' singsong voices cut deep.

Tears flowed freely. It was hard to catch my breath.

I failed... So many are gone... I failed. I choked out a cough between sobs. It was an impossible situation. The odds were against me. The four other dots near me sobbed as well. It was hard to focus on the four that made it out.

A tussle below caused the ground to shake. I stood up and looked down at the camp. The two smaller gnashers fought over a satchel. It slipped from one of their hands. The ground shook as they both tumbled. The slightly larger gnasher rolled over one of the bags.

I shrieked. My hand flew over my mouth.

My chest felt tight, like I'd swallowed a stone. My hands shook against the rough leaves of the bush. The numbers still echoed in my ears—Sixteen eight two. Twenty-seven nine six one. They were gone. So many of them. Because of me. My knees buckled, and I sank into the dirt, pressing my forehead against my hands.

One of the four dots clutched a scrap of fabric, their knuckles pale around it.

Another dot, skinnier than the rest, stared blankly ahead.

None of them spoke. None of them moved.

I tried to form words, something—anything—to break the silence.

"I'm... I'm so sorry." I muttered.

My voice cracked, and the words felt useless. The dots stared at me—eyes hollow, faces pale. The situation was beyond my ability to repair. I dug into my satchel and pulled out a handful of berries. Despite the emotional tremor in my legs, I forced myself to stand. The ground seemed unsteady beneath my feet. With a watery attempt at a smile, I extended my hand, offering the sweet-smelling berries.

"Eat. Please."

One by one, they took them. Slow. Hesitant. But they

took them.

I crouched down to their sitting level, meeting their eyes as best I could.

"I know I failed you today. But I swear—I won't stop trying. I won't stop until we're all safe."

My eyes burned and felt puffy.

It's alright to feel. I reminded myself. *Let the tragic moment pass naturally.* I cried until the rise and fall of my chest normalized.

"Come with me," I said. "I have a safe place to stay."

Each of them stood bravely and with determination on their faces. The walk back was quiet. It was clear the experience had shaken us all. All I could do was place one foot in front of the other until we reached the grove.

VI

The Caretaker

Eulalia

The four dots rested while Al and I worked in the fields together. I filled him in on the experience.

"What a tough first day." He sympathized.

"It feels impossible!" I said.

Al stopped pulling weeds and looked at me. Sunlight glittered off his brown eyes. He stared at me with a gentle intensity and said, "Had someone told me yesterday you would save me, I never would have believed them. Yet, here you are. And you did just that. If there was ever a time for the impossible to be possible, I think it's now."

He gestured to the grove and looked for the top of the

trees.

"I think I know what to call this place." He said in a tone of encouragement.

I finished pulling a weed and stood next to him. I smiled and looked around at the peaceful evening.

"Countable Grove," He suggested, with a smile.

I tried the name out. It helped me feel a new sense of home. Like this was a place worth fighting for. A place to encourage others to be themselves. Somewhere outside the influence of Arithma's grasp. Safe.

"Countable," I said, "Because none will be forgotten."

"Come with me," he said, picking up a large rock. He made his way towards the ancient hollow and placed the rock at the tree's base. Then another and another.

"How many dots did we lose to the gnashers?" He asked.

"They crushed 25. And we lost 36 in total." I responded.

The four dots I saved gathered around, curious about the commotion.

"We suffered a tragic loss today," Al continued. "Counting allows us to recognize when we have lost and those that are missing. Today, I place a stone for each dot we have lost at the base of this tree."

Tears flowed down my face, glistening in the sunlight.

"Thank you," I whispered.

The weight of their absence lingered but was somehow lighter. I reached for a smooth stone, feeling its coolness against my palms, and gently nestled it among the others. A sense of release washed over me, as if the burden of remembrance had lifted, carried away by the gentle breeze rustling the leaves above.

"Though they were unnamed," I said between sobs. "We knew them. We recognize them because we can count them."

The four dots stood in silence as Al and I placed the stones. Thirty-six in total.

Al stood next to the pile and said, "We will place a stone here for any that depart. When they come back, we will remove the stone. This will ensure no one will ever be forgotten."

A few hours later, a newfound determination surged within me. I was eager to glean lessons from the past and shield myself from further losses. The warm candlelight danced across the Aether's Almanac promising future success. The heavy book rested on the table. Its thick leather binding creaked as I opened it. I ran my thumb across the worn pages, searching for details on the other dangerous creatures that lurked in our area, the musty smell of old paper filling my nostrils.

Ellis, one of the rescued dots, sat in a recessed part of the wall. Her fingers tracing the cool, smooth surface of the amethyst gemstone. The stones' purple hues shimmering faintly in the dim light.

I turned a page.

Ellis placed the gemstone in her pocket and silently observed my movements.

Ronan noted her shift in attention and followed her gaze. These two were the first ones up the hill earlier today. Thorne and Archy, the other two, were already asleep.

"Do you know what those creatures were called?" Ronan asked.

"Gnashers," I responded. "They lure their enemies in with sweet-smelling perfumes and then capture them forever in their bags."

I stopped on a page titled *The Gnashers Curse*.

"Come here," I invited.

Eager to do something, they were both quickly at my side.

I pointed to the worn-out page as I summarized the contents.

"This is a curse that creates gnashers."

Their eyes fell on the illustration. Dots of all sizes stood in a circle, each one being pulled into different parts of the creature.

"It's a form of combining magic," I said. "Performing it with a partialed dot in the group forever combines them into a single dot known as a gnasher. The cursed form searches the island, seeking to be made whole again."

"Why would any dot want to merge like that?" Ronan asked, making a strange face.

"I don't know that anyone sets out to merge as a fraction. Typically, the spell combined several whole dots to create a larger, more powerful single dot. The combining spell would enable them to fight back against a gnasher. Sometimes things go wrong, and partialed dots get caught in the mix."

"That seems stupid!" Ellis said in frustration. "To fight a gnasher, we need to risk becoming a gnasher ourselves?"

"Most insightful!" I said, genuinely impressed. "We can fight back in other ways. The merging spell has too many ways it could go wrong."

I turned the page.

"In ancient times, merging was the only way for dots to fight back. The witches that performed this spell were called Additioneers. They would combine, steal the satchels, and then recruit the dots they saved to their cause."

They looked at each other as if trying to imagine what it would be like merging into a single dot.

Ronan chuckled nervously.

Ellis grew bright red.

"Don't make it weird," I said with a hearty laugh.

Ellis's look of disgust faded to a chuckle.

Ronan eventually gave in to the laughter as well.

The tension faded. It was a relief to laugh. It felt good.

"We've got a long day ahead of us tomorrow."

Al said, placing a basket of berries on the table.

"You're right," I responded. "Let's all get some sleep."

A flurry of activity filled the next few days. Every time I went out, Ellis and Ronan insisted on helping. For their safety, I encouraged them to stay within Countable Grove and familiarize themselves with the surrounding hazards. They became determined to learn as much as they could to one day join me.

The grove felt the effect of Al's patient guidance. Each dot I saved became a mentor under Al's careful watch. Despite the overwhelming number of dots that joined our ranks, Al's smile encouraged each to contribute in their own way.

The dots I found were often in perilous predicaments. Mangled by a bear, trapped inside a deep cavern, or unable to move because of hunger and fatigue. Occasionally, I would come across a group of dots flourishing. Though hesitant, they would usually agree to join us.

The passing weeks showed significant progress. The once quiet grove was becoming a bustling and busy place. Purposefully, the five original dots helped all acclimate to the new environment. Their diligent efforts filled the grove with a sense of community.

In the bustling grove, the chatter of dots filled the air as they eagerly discussed my legendary adventures. Their anticipation was palpable. They longed for my return to listen to more tales, which they would embellish and pass on as weeks passed.

"I was there that day," the newcomers would eventually claim, their voices ringing with the conviction of firsthand experience.

"The beast was four times their normal size, with fangs as long as my arm." They exaggerated.

Ellis and Ronan paid little attention to the gossip. Ellis was too busy studying the almanac or creating runes. Ronan was setting up obstacle courses to keep everyone in shape. They proceeded like this for weeks.

"I'm ready!" Ellis insisted. "I've made my own runes and have read every page of the almanac."

Ronan stood in a stoic pose beside her. His muscular arms folded.

"I'm strong and agile." He insisted, "I can get to places that may be hard for you to reach."

Al gave me an encouraging glance.

"Alright," I relented.

Ronan pumped his fist in the air.

Ellis bounced with delight.

I wanted to ensure their first day out was memorable, but safe. A few days prior, I saw an abandoned gnashers' camp. With just me, it was too dangerous. With two others, we had a fighting chance.

The three of us crouched in a bush.

The gnashers' satchels rested like boulders on the opposite side of the clearing. Bones littered the ground.

"Where are the gnashers?" Ronan asked.

"It's like this every time I come by," I said. "It's a rare situation that could go wrong quickly."

"It smells like forest," Ellis noted. "Where's the

gnashers' perfume?"

"Great perception," I responded, proud she had noticed the obvious concern. "That's why I don't think gnashers were involved in this setup. I'll go in first. You two hang back. Any sign of trouble come help."

They gave me a look as though about to argue.

"Fine," Ellis said in disappointment.

I walked out into the clearing, expecting the trap to spring. Nothing happened.

The first bag was empty. I looked for the opening of the second satchel.

Footprints, lots of them, led to the forest on the other side of the clearing. I remained focused on the bags.

We'll find them next, I thought.

I opened the second bag. Hundreds of dots were inside.

"This way quickly!" I said.

Ronan and Ellis were standing near the perimeter of the forest waving them towards them. They provided them with food and water.

Another bag was empty, and the other two had hundreds more. We provided for them and headed back to the grove.

"That felt too easy," I said to Ronan and Ellis. I didn't mention the footsteps. Whatever led the dots away was the obvious danger.

The moonlight bathed the scene in a soft glow, casting shadows that danced around us as we returned to the grove.

"You are getting good at this," Al said with a smile.

Within a few moments, I stood in front of thousands of dots. Tired of my stories being retold, I turned to Ellis.

"Let's have you tell this one," I said.

"Today, we celebrate Eulalia's efforts," Ellis shouted, pointing in my direction.

I stared in confusion. That's not how I expected her to start.

They all cheered and formed a tight circle around us.

Ellis continued, "She has saved us from a danger we couldn't even imagine. She works tirelessly to save others and prioritizes their needs over hers. She is a loyal friend, and I am grateful for all her efforts."

Ellis turned to me.

"We made you a gift!" She chirped.

Thorne brought out a grass-woven fabric cloak. They intentionally faded the green and dyed it dark blue to imitate royal robes.

Al joined us in the circle and shared, "We made this cloak to keep you safe as you continue your search for lost dots."

Thorne extended the cloak towards me. The fabric felt soft against my fingertips. A heavy weight settled on my chest, a mix of gratitude and unworthiness. The myths of my journeys weighed heavily on me. Despite my efforts, I couldn't help but feel I was responsible for everyone in danger. While they fastened the cloak, its exquisite detail captivated me.

A pocket for the knife, making it ready for action. Plenty of pockets for gems, jewels, and berries. I smiled. It was nice to be appreciated.

"I think you've earned this," Al whisper. "You've shown that you have changed. I will stand by your side."

He pulled me into a hug. His warm embrace melted away my feelings of unworthiness. I hugged him back, fighting tears.

"Allow me to introduce the legendary Caretaker," Ellis announced, gesturing towards me.

The crowd erupted with cheers.

Al continued, "We give you this title out of respect and love. A title that will help others quickly identify your goodness. Eulalia has been reborn into a more loving individual."

I was speechless; tears of joy filled my eyes. Then, from the depths of my fears, a thought emerged: *Arithma is still out there.*

Feeling a horrible omen looming over the grove, I wasn't sure how to respond, so I settled for, "Thank you."

VII

The Five Domains

Eulalia

The next morning, I had Ronan and Ellis look for dots in a safer region of the forest. This enabled me to return, alone, to track the footprints leading away from the gnashers' satchels. The mass number of dots broke the forest underbrush, leaving an obvious trail. I followed the prints to an abandoned campsite. The smell of burned pine was still in the air.

The camp had an eerie atmosphere. Shadows danced ominously from the recently cut tree stumps. Patches of jagged, thorny vines littered the ground—bindings. Each pair of vines showed the imprint of something kneeling.

Prisoners. I thought. The vines were to keep the dots from running away. Irrationals are the only dots with the power to command nature. *Did Thrall do this?*

The thought sent a chill up my spine. Arithma was cruel, but this was beyond anything I had seen. If it was Thrall, Arithma gave the order. I looked around to see where they might have gone. An enormous pile of logs blocked the exit on the far side of the clearing. All footprints led to the center of the pile.

I gasped.

Were they crushed?

Using reduction magic, I carefully removed a few logs. Loud wooden thuds echoed. The earthy smell of damp wood filled the air as the logs shifted to fill the gaps. The strain of manipulating such massive objects drained me, my knees trembling with exertion. With a heavy breath, I made two more logs vanish. Pushing through the fatigue, I continued until I could finally see through to the other side. Single-file footprints brought a glimmer of hope. Maybe this was just to prevent others from following.

I recovered my strength and returned to Countable Grove.

Over the coming months, I frequently returned to the path and tracked it deeper into the forest. With each return, an obstacle would block my progress. After the removal, I had to stop and recover my strength. On every recovery visit to Countable Grove, Ronan and Ellis shared their stories of success. I knew this path led to something important, something critical. It eased my mind to know others were saving dots. With their help, our numbers grew to hundreds of thousands.

Al encouraged each dot to find their own way of contributing. All dots had their needs met, regardless of their contributions. Countable Grove was bustling and unstoppable—regardless of my involvement.

Arithma taught me that dots would fall into chaos without someone controlling them. Yet, I was actively experiencing a direct contradiction to the belief. Letting others live authentically wasn't chaotic; It was self-organizing. This was an achievement Arithma could never access. This type of order requires letting go of control. She could never do that. Her system requires her to be at the top. It requires her to be the order.

The next day, I found the end of the mysterious path. Resting near Origin Tower was a cliff. A boulder the size of a gnasher sat at the base of the cliff. A powerful force had split the rock down the center. The sharp, irregular edges jutted out. Their cold, unforgiving texture spoke of a precise, unnatural break.

A quiet shriek escaped my lips as I threw myself behind a nearby tree.

Thrall, Arithma's henchman, was moving through the open space with a hasty sense of purpose.

I glanced around, looking for any sign of Arithma. She was not anywhere I could detect. I patiently observed Thrall moved about the space until well into the afternoon.

He looked as though he was making preparations for something.

I heard a whisper from beyond. Not from an existing presence. Something in the depths of aether itself. It was a call to action.

"The Cursed Dragon," the voice said.

My heart beat faster. The already creepy vibe turned dark as the realization settled in.

Arithma would never cast that spell. I thought. *She would never be so careless... Would she?*

An ancient prophecy foretells the dragon's freedom, followed by the destruction of all life on the island. The dragon is the destroyer of logic and aether itself.

I always thought Arithma wanted peace in her own cruel view of the world. This seemed out of character, even for her. To awaken a force like this would be the demise of all of us.

There has to be something I'm not seeing yet.

Thrall wandered into the forest, allowing me a chance to prevent the spell. I ran as quickly and quietly as I could to the split rock.

My heart sank at the sight of a motionless dot resting inside. Completely uncovered, I quickly draped them in my cloak.

Bruises and shallow cuts marked the dot's face. They had a faint pulse. The naming curse prevented them from moving on from this life. I placed their head gently on my lap. With hasty movements, I pulled a few plump berries from my cloak and held them above their open mouth. My hands compressed, releasing a sweet scent. The cool juice dripped from the palms of my hands.

My eyes darted from each jagged shadow of the interior, terrified of getting caught.

A dark cave entrance into the cliff loomed on the other side of the rock.

An icy dread settled in my stomach as I shivered, sure that something was observing my every move from the shadows.

The broken boulder was full of opal gems. I placed one of the larger pieces in a pocket and carefully placed the dot over my shoulders. Moving swiftly, I made it to the edge of the forest.

I crafted a stretcher out of nearby sticks and large leaves to ensure I didn't hurt the dot further. I secured the dot to the makeshift bed and dragged them home. I tried to make the journey as comfortable as possible and winced as the bed fell off some of the larger rocks and roots. The last traces of sunlight highlighted the treetops to the entrance to Countable Grove.

"Almost home," I said, in hopes they could hear me.

The innkeeper's hollow is near the entrance for quick medical attention. The innkeepers are our healers who care for the rescued ones. They support newcomers and aid in their recovery. I made my way inside.

"Welcome back!" Ronan said, stepping into the room. His cheerful tone faded as his eyes landed on the motionless dot.

Archy—the leader of the innkeepers—froze mid-step. A clay mug slipped from her hands, shattering against the wooden floor. The earthy scent of herbal tea filled the air as liquid spread across the planks.

"Oh, no... what happened to them?" she asked, her voice trembling with concern.

"It's an ancient dark magic. I think she will survive," I said, my voice strained under the exhaustion.

Archy stepped closer, her brow furrowed as she examined the dot's injuries.

She said, "The naming curse becomes more terrifying every day. I consider what it means for dots in Countable Grove. What other dots remain hidden from view and cannot die? Let's get her settled—Ronan, grab those clean linens. We'll need warm water and a way to extract berry juice."

Ronan snapped into action, the lightness in his usual demeanor replaced with focused determination.

"They'll be okay, right?" he asked, apparent concern in his voice.

"We'll do everything we can," Archy said firmly.

I gently laid the dot down on a bed of soft moss and tucked a blanket around her fragile form.

Archy knelt beside her, already grinding berries into a juice.

As candlelight flickered against the wooden walls. I let out a breath I didn't realize I was holding.

"She's safe now," I said, almost to myself.

Archy glanced up, her eyes soft but resolute.

"And she's not alone anymore."

Ronan set the last of the supplies down beside Archy and gave me a small, reassuring nod. "Whatever happens, we've got her now."

Archy opened the dot's mouth and helped them drink some liquid. Life returned to their body, bruises faded to a greenish tint, and the cuts began to scab over, but she remained motionless.

Concerned about the fate of the entire island, I returned to my personal hollow. The book from Arithma's cursed chambers rested on a tea table. I flipped through the pages until I found a dark image of a broken boulder releasing the spirit of a creature long forgotten. The creature flew forth, devouring the whole of the island. Dots from each of the five domains ran from its jaws.

The Cursed Dragon.

My heart sank. The scene was surreal and distant, and yet somehow familiar. I felt disconnected and hopeful that it wasn't true. Regardless, it would be unwise to ignore the evidence before me.

"Arithma attempted to raise the beast." I said in a whisper.

Hoping for answers, I read the description,

A spell born from the depths of deception. A curse none will cure. Aether is born of consistency; curses spring forth where consistency fails. The Cursed Dragon is born from an inconsistency that has no remedy. It will be the undoing of aether itself.

Laughter erupted just outside my room, making me jump.

Ellis was telling Al about a funny rescue she had made.

I tried to block out their voices and continued to read.

I, the author, do not think it is responsible to include the ingredients of the curse, nor do I know them. Instead, I provide a potential cure.

My eyes widened.

Buried in the depths of hope lives a chance to unite the five domains. Fighting off the dangers of the beast requires courage, diligence, and the assembly of a council. A council made by the binding magic of Al-Khwarizmi himself. A magic that unites the five domains.

"Eulalia! Find anything interesting in the almanac?" Ellis's voice rang out from the entrance of my study.

Dazed and overwhelmed, I turned the book to her.

"See for yourself." I said with glazed over eyes.

She glanced at the page.

"No!" she said. "There's no way!"

"The path I've been tracking for months leads to a place where Arithma is preparing to wake the beast."

Her eyes stared at the page.

She said, "I've read this page so many times, wondering

what the beast would be like. I never thought we would need to fight against it."

She turned and asked me, "How do we fight back?"

"We have no choice but to visit the five domains and try to assemble the council."

"How?" she asked sharply. "I've never seen dots from the other domains. Do they even exist?"

"That is because you are not looking in the right places. They used to exist all around us, but Gaia, Arithma's predecessor, chased them out of Countable Forest. Luckily, the book mentions where to find them."

I pulled the book close and pointed as I read.

"Any hope of survival will require the five domains to unite under a single spell. Aether in its purest form is a counting magic—naturals require this counting to survive. Negative dots live below—born from the wells of the additive inverse. The rational domain exists among naturals and negatives. However, some consider themselves pure. They dwell in the mountains and live off the rational fruit trees provided by the core of the mountain. As rationals, negatives, and naturals exist within the countable aether, irrationals exist where countable aether does not. Little is known about this mysterious, uncountable aether, but it seems to infuse these dots with immense power. Complex dots live in the sea beyond. These dots are a combination of countable and uncountable aether."

"So it's settled," Ellis said with determination. "We will

unite the five domains and fight back against The Cursed Dragon."

"Indeed, we will." I agreed. "First, we need to return to the split boulder. If Arithma has not completed the spell, we will save ourselves a great deal of effort."

With a bow, Ellis said, "I will get Ronan and we will make preparations."

My mind reflected on the condition of the dot I rescued. I feared the spell was already complete. To protect them, I visited the blacksmith.

We worked tirelessly through the night to craft a necklace. The opal gemstone I took from the split rock would have unique magical properties. Large spells pull from natural aether and infuse nearby minerals with their energy, making them perfect conduits of magic.

Letting the wearer decide what runes to place upon it was customary, but I broke tradition. I felt they needed more protection than the average dot. On the back, I placed a vertical and horizontal cut. On the front, I carved an inverse symbol.

"You didn't make me a necklace!" Ronan said, nudging Ellis with an exaggerated pout.

I raised an eyebrow and replied flatly, "You're not unconscious."

"I usually prefer Ronan unconscious," Ellis quipped,

crossing her arms with a smirk.

Laughter rippled through the group, a brief but much needed lightness after days of worry. I let their humor fade naturally as I gently set the necklace beside the injured dot.

To my relief, the dot stirred. Their eyelids fluttered. With visible effort, they scanned the room, their frail chest rising and falling in shallow breaths.

"I'm The Caretaker," I said gently, leaning forward to meet their uncertain gaze. "You're safe now, but I found you in a cavern. Do you remember what happened?"

They shook their head weakly. "I... I remember nothing. Where am I?"

"You're in Countable Grove." Ellis said in a warm voice. "It's a safe place, a home for those lost or forgotten. You can stay as long as you need."

Archy held out some water in a wooden cup.

The dot took a cautious sip, their hands trembling slightly as they gripped the cup.

"Do you have a name?" Archy asked.

"I don't... I don't think so."

I picked up the opal necklace from the stool and offered it to them. The gem caught the candlelight, glinting softly in the dim room.

"We found you surrounded by stones like this," I said. "How do you feel about the name Opal?"

The dot's eyes lingered on the gem, their lips curved into a faint smile.

"Opal... yes. I like that."

"Welcome to Countable Grove, Opal!" Ellis said encouragingly, her voice carried the weight of sincerity.

I gestured to the innkeepers standing nearby.

"These dots will help you settle in and ensure you get the care you need."

Satisfied that Opal was in safe hands, I rose to my feet. Ellis, and Ronan followed me as I stepped outside into the cool night air.

"Get some rest. We will leave at first light." I said.

The early morning gray dissipated as the sun added color to the sky. I stood at the edge of Countable Grove, gut churning in anticipation. The warm hues dancing in the sky felt like a betrayal.

Why doesn't the sky reflect the dangers ahead? I wondered.

Deep in my gut, I already knew the truth. Arithma woke the dragon, but we needed concrete evidence. This type of thing required certainty. Thrall was making preparations yesterday. Hopefully that means there's still a chance to disrupt.

Ronan and Ellis gave an encouraging nod as we stepped into the grass field beyond. We rushed through the forest and past each of the removed barriers until we stood before the ominous scene.

The three of us hid inside a bush to observe quietly.

A hooded figure, their breath misted in the cold air,

eased their hand across the boulders' jagged surface. The early morning shadows provided cover as they moved skillfully.

"They're looking for opal," Ellis whispered.

Something felt off. A tight knot of dread settled in my stomach. I glanced at Ellis and Ronan and whispered, "Let's see if we can follow them."

But when I turned back, the figure was gone. Not a sound, not a single trace of movement remained. The clearing was empty.

Moments passed.

Nervously, we stepped out of the thicket. Every twig snap and rustle of leaves put me on edge. Inside the cavern, Opal's imprint still marked the cold stone floor. My own footprints circled where I had knelt, but the hooded figure left no tracks. It was as if they had never touched the ground—or as if they weighed nothing at all.

"What are we looking for?" Ronan whispered, his voice tight with unease.

"Anything that—" I started, but Ellis's scream cut me off.

I spun around, my heart in my throat.

Ellis was gone.

Before I could react, the cloaked figure stepped from the shadows, emerging with an unnatural stillness. Their face remained hidden, swallowed by the dark folds of their hood. The heavy cloak draped to the ground, unmoving despite the slight breeze.

With a flick of their hand, Ronan released a startled yelp—his form shrank until he was no larger than a berry. He landed beside Ellis. Their tiny figures were barely visible against the cavern floor, waving their arms high over their heads.

Relief flowed over me—they were still alive.

I crouched low, scooping them gently into my palm.

The hooded figure ominously observed my movement.

I kicked the dirt and cast a duplication spell.

The tiny dust particles fluttered to life as the space filled with dust.

I ran towards the forest, the weight of my friends pressing into my hand, my breath ragged and sharp.

I can reverse this with Al's ring. I thought.

The trees loomed ahead, dark and protective, their branches reaching out like arms ready to shield me.

Almost there—CRACK!

A deafening sound rang out, and everything went dark.

VIII

A New Friend

Opal

Thunder rumbled me awake, dragging distant, fearful memories to the surface. I recalled the musty scent of the cave mixing with fresh rain. The warm, cozy hollow of Countable Grove provided a shell of protection. Yet its emptiness echoed in my mind. I was safe, but I wasn't home. Something was missing.

Dots here are welcoming and friendly. Over the last few days, they've taught me the basics of survival. They've provided me with a place to stay. They call me Opal. A name that feels like someone else.

Deep in thought, I crawled out of bed and stepped onto

the ledge outside my hollow. The cool drops of the rain trickled on my skin. I held out my arms, feeling a greater connection to the dark emptiness of the night sky.

A new sense filled my mind. I felt connected to the rain, the storm. Like I was hovering above the ground looking down on a small village. Each drop brought attention to a new section of the grove. The hollows carved into the sturdy trees provided individual dwellings for each resident. They all looked mostly the same but had hints of personal expressions. There were thousands of them. Dots slumbered peacefully, tucked in their warm blankets made of woven grass.

My body shivered. I turned to grab the blanket resting on my bed. The water flowed around the opening of the hollow. A minor cut allowed drinking water to pool inside the room without causing damage. I dipped my hands in and took a drink.

Warmly wrapped in my blanket, I continued to watch the dark forest. My place was near the top. They were making more hollows above mine. The builders would work on it during the day. Watchers and defenders would come out to protect us at night. They were a lively bunch and always happy to see another dot. Isolation is common for them.

I stepped into the rain again and took a deep breath. Fresh and calming. With my eyes closed, I tilted my head back. The drops felt cold on my already chilled face. Rainfall echoed in the treetops above. I opened my eyes to watch the drops falling towards me. The rhythmic sound of the rainfall was almost countable, but just out of arm's reach to keep me in mystery.

Unaware of how slick the wooden floor had become, I stepped back.

A sudden ear-splitting shriek shattered the quiet.

My foot slipped.

I fell into the drinking water. Deeper than I realized, the current pulled me with tremendous force. I grabbed hold of the edge. I tried to fight. It was useless. The water was too powerful.

My arms slipped, and I flowed down the tree. In a blur, I hit the dry ground with a thud.

Gasping for breath, I coughed and wheezed. Grateful I somehow survived. I looked around.

I was in another hollow... Or maybe my own?

I sat up to take in the space. My head ached in pain. The wall above the bed had leaves hung to dry, confirming I was in someone else's room.

Still dark, the forest looked different from where I was now. I searched for the source of the shrill noise. There were brief and distant shapes in the shadows.

"Goodness me!" A calm voice from behind me said, "In all my time in this grove, I do not think I have ever seen someone try to ride the drainage."

There was a hint of admiration in their voice. They joined me and looked out over the grove.

"Are you alright?" they asked.

The question surprised me. I became trapped in my mind.

Of course, I'm not alright. I thought but couldn't find the strength to say it out loud. *I just fell. What was that shrieking noise? What are those shapes?*

I glanced to see who was talking to me. The surprising events of the evening were just the beginning. Standing next to me was a half dot—as in half my size.

Dots were all the same size—or so I thought. This specific dot was older than me. The mug she held looked oversized and disproportional to her tiny hand. Heavy reading classes dug into her enormous nose. Her caring eyes beamed at me, magnified by the reading glasses.

"You're safe here." She said after a few moments of quiet. "You're welcome to stay as long as you'd like. I don't get many visitors these days."

She took a sip of the mug.

"Long ago, we didn't have dedicated individual dwellings. We just stayed wherever it was convenient. Most homes were overflowing. Eventually, there were privacy complaints, and *poof* almost overnight, everyone had their own place."

She chuckled to herself.

A large squirrel rushed into the hollow and startled me. It shook off the rain and pranced over to the stranger. She brushed her free hand against the animal's fur.

"No means for alarm," she said reassuringly. "This is Wriggles. Created from ancient magic. It's a fascinating story."

She placed down the mug and scratched the squirrel's head.

In response, Wriggles leaned into the head scratches. Its eyes partially and lazily opened. It moved its body as if asking her to scratch in specific regions.

She quickly and willingly complied.

Usually, we would try to keep animals out of the grove. They eat our crops and cause problems. This one was a pet for her, though. The large animal walked to the back room and curled into a ball.

"I've never seen a squirrel up close before," I said in a burst of nerves. "I didn't realize they were so large."

She smiled warmly as if delighted to hear me talk. Making no note of my quavering voice, she said,

"This one is unique. Common squirrels are about the size of your arm. This one is so large because of that ancient magic."

She let out a reminiscent half laugh and with a crooked smile, she glanced at Wriggles.

"They deserve it, though." She said playfully, "Such a helpful creature. Wriggles pulled you out of the drainage tonight."

All at once, things made sense.

"Thank you, Wriggles," I said in my bravest voice.

I stood up, and my head throbbed in response.

The fear of the distant nightmares faded, along with my desire to be somewhere else. It felt as though I had known this dot my whole life.

That's silly. I thought. *I don't even know her name.*

"My name is Opal," I offered, questioning whether it was false.

"Opal's such a beautiful name." She responded in admiration. "The Opal gemstone is a sturdy and reliable source of magic. I'd imagine there is more to you than meets the eye."

Opal. A gemstone, a source of magic. Could my name mean something more? This dot made it sound important, not just a label, but a purpose waiting to be discovered.

"Come, let's get you some tea!" she said and walked behind the dividing wall between the entry and the back rooms.

The thunderstorm was coming to a close.

The loud noise must have been lighting. I thought, trying to convince myself I wasn't crazy.

Regardless, I was safe. Each step deeper into the hollow caused my head to throb in protest. Furniture skirted the walls. Wriggles, my savior, was curled up in a nook carved into the furthest wall. A sturdy and well-kept bookshelf rested opposite to the furniture.

The mysterious half-dot in the kitchen was having difficulty reaching the top shelves.

"Is this tree growing?" She asked. "Do you mind helping me get the jar from the top shelf?"

There was a small fire under a metal pot full of water.

The shelf she pointed to had several jars with labels.

Jam. Jelly. I reached up and grabbed the jar labeled tea leaves.

"Is this what you needed?" I asked.

"Thank you!" She said delightfully, lifting the lid and placing a few leaves in the boiling water. "Have a seat, dear. I'll have the tea ready in a moment."

An open archway separated the two spaces. I walked over to the bookshelf and glanced at some titles. *Teas Around Arithmetic Island. Edible Plants That Might Surprise You. Primes, An Adventure Into The Unknown.* The same author, Tia, wrote these three books.

I took a step back to glance at some of the other titles. *Brotherhood Vibrations*, by Gregg. *Aether Analysis: A Magic From Within*, by Dag. I grabbed the book titled *Logic and Reasoning*, written by Zen.

I collapsed into the chair furthest from Wriggles, still nervous around him. A few books rested on a coffee table in the center. The space showed the wear of countless hours of reading. My body sank further into the seat. The weight of the book offered a calm reassurance. Warm candlelight caused my eyelids to grow heavy.

I jumped as pots and pans banged together. A few moments later, the dot returned.

"I'm Tia, by the way." She said, offering me one of two large mugs. "Ronan found me living off tea leaves in the wild. He called me Tia as a joke. It caught on."

I collected the mug from her. It fit nicely in my hand and seemed proportional to my size.

"You mean you didn't choose your name?" I asked.

"Nope," she responded in a matter-of-fact tone. "It took some time for me to realize, but I have a significant connection to tea. I love to study and find it delicious. It's got healing properties and enables enhanced focus."

With her words, my name took on a different weight.

Was I destined to love Opal as much as Tia loves tea? I imagined placing an opal gemstone in a pot of hot water. It was ridiculous, but I found it difficult to picture the reality of loving opal as much as Tia loves tea.

"Thank you," I said. "You've been incredibly kind."

I raised the cup to my lips. The warm liquid swirled inside. I took a sip. A bitter taste filled my mouth, reminiscent of dirt. I quickly spit it back into the mug. The lingering flavor clung stubbornly to my tongue.

Tia laughed heartily.

She said, "That's why so many eat the berries instead of drinking the tea. The berries are sweet. Tea is an acquired taste."

I placed the mug on the table before me and sat back.

"You may find the tea is helpful in easing that headache of yours. Even with the little sip you took. It's powerful stuff."

She was right. Within a few moments, my headache faded to a dull annoyance. The weight of the book was comforting, resting in my lap. Warmth radiated from the grass fibers of the chair pressed against my back.

Tia continued to talk.

"Tea is a special thing. It's part of nature and delivers nutrients directly to the body. When I first got to Countable Grove, I would make large batches. Everyone would come around and enjoy tea."

Her voice was soothing. My eyes were getting heavy.

"Unfortunately, the aroma attracted a few hawks. That put a quick end to the large production of tea. Now, it's mostly made in individual homes. There are only a few of us around that still enjoy tea on the regular."

The warm light offered a particular safety from the storm. Wriggles, on the far side of the room, vibrated the furniture with his rhythmic snores.

Tia continued, "The Caretaker usually comes over for a pot of tea now and then. She's always got a new story to tell. The name Caretaker suits her. She is a wonderful dot."

Tia placed a large blanket over the top of me. I gave into the weight of my eyelids. Over the next few moments, I would come in and out of hearing what Tia was saying.

"Seems like a waste to only harvest the berries. I have a lot of fond memories of the days before Ronan found me. There was a patch of grass with a single tree. I stayed under the tree during the day and then would brew the leaves at night."

Sleep caught up with me.

Keeping my eyes sealed shut, I lay relaxed. A muffled

noise filled the air. I pulled the blanket off my ears and opened my eyes.

Tia was still talking.

Warm sunlight beamed into the room. It was late morning or early afternoon.

How long had I been asleep? I wondered.

Her voice, still soothing, showed no sign that she knew I had fallen asleep.

"Have you been talking all night?" I interrupted her.

Cup of tea in her hand, Tia looked surprised, as if she hadn't noticed the obvious change of light in the room.

"Oh dear! You're right. We've talked throughout the night. I hope you won't be too tired for training today. I will get you a fresh cup of tea."

"Oh, right! Training!" I said, rolling off the chair.

My body hit the ground with a thud. Both my arms were numb—one from the pain of the fall and the other from waking up after sleeping on it strangely. After the pain subsided, I thanked Tia and stood up to leave. Before I made it to the door, Tia stuffed another cup of tea in my hand.

"No need to return it, dear. I find most dots do not have mugs for tea. Please come back sometime."

Her words were genuine and loving. I paused for a moment at the entrance, not wanting to leave. The empty feeling I had last night was gone. Living alone was uncomfortable. Tia made things better. Easier. I just stayed the night here, but it felt like something more significant than that. It felt like a home. Like I belong here with Tia.

I made a mental note to come back after training. The thought made me smile—something to look forward to. With a deep breath of cool air, I stepped out.

IX

Chaos in Countable Grove

Opal

At the base of the tree, the local store owner was half their usual size. He attempted to grab a bag that rested on a post.

"Hilarious. Who moved my bag this high?" They snarled. Other workers of varying sizes appeared nervous as they scattered into the shop.

I pulled the bag off the post and handed it to the dot. They angrily jerked it out of my hands.

"Thanks," he muttered. He then disappeared into the back of the shop.

I shrugged off the event and attempted another sip of

the tea, which still tasted horrible. I poured out the contents into a nearby bush. The healer's dwellings served as a landmark for getting around in the bustle of work in the grove. The training grounds were the large vined post to the north of them.

My previous commutes were nothing like today. The grove floor was full of a new form of diversity—dots of varying sizes. I could easily see where I was going, with no need to jump. There were only a few others as tall as me.

"Watch where you're going!" growled a knee-sized dot I nearly plowed over.

"Oh, sorry," I responded. "I didn't see you there."

"How rude," said another of the same size. The two dots were part of a group of six that looked identical. All of them pressed through the crowd aggressively.

To my right were two identical dots that wore the same backpacks and clothes and skipped in sync.

Has it always been like this? I thought, attempting to recall the past few days.

I was confident that yesterday, everyone around me was the same height. Last night, Tia attempted to reach the jars on the top shelf—*could she always reach them? Someone had to put the jars up that high. It was almost as if she had shrunk overnight.*

I shrugged off the thought and proceeded to the training grounds. When I arrived, the whole place was empty. I moved to the nearest wooden table and placed Tia's mug down.

"Anyone here?" I shouted.

"Opal!" a voice rang out.

A third of their normal size, Aleph stepped out from behind a healthy berry bush. He animatedly carried his dead potted plant toward a nearby desk.

Aleph was my only friend in class. Despite not being very good at training, he lived for it. His eagerness to learn, and his fear of missing out, made it hard for him to concentrate on one thing.

He tripped over one of the protruding roots, upturning the pot in his hands. We both looked down. Soil covered the floor.

I ran to his side.

"Aleph? Is that you?" I asked.

"Of course it's me!" His voice was higher pitched than usual. He gave me a look as if I just made a ridiculous joke. Another dot entered the room. They looked, dressed, and moved just like Aleph.

"They canceled training," the second dot declared, sounding exactly like Aleph.

The two Aleph replicas stood before me, bubbling with energy and excitement.

I helped the Aleph that tripped get up.

He brushed the dirt off his shirt.

"It's nice to see you." Both of them said.

"Why are there two of you?" I asked.

They exchanged a quick, bewildered glance. Startled, they both stepped back, their movements accompanied by a gasp.

"Who are you?" they questioned in unison, the words echoing in the air. Slowly, they circled each other, their eyes scanning the identical cuts and bruises on their bodies.

Turning towards me, they inquired, "Why are you so large?"

"I think I'm normal size. You are smaller than usual."

The Aleph on the left scoffed at the notion.

How could they not notice they were smaller? I considered if I would notice if my height had changed. *Perhaps I'm the one getting larger?* I pondered the thought briefly.

The two began fighting over a book. Unsure of what else to do, I tried to change the subject.

"Do you know why they canceled training today?" I asked.

The words had no effect as the two continued to argue.

"You stole it! Give it back!" One of them demanded, picking it up.

"How can I steal what is mine?" The other said, taking hold of the book.

"Quiet!" I demanded. They both stopped and swiveled toward me.

One of them let go, looking innocent.

"Sorry, I didn't know I was such a crook." He said, gesturing to the other who had the book in their hands.

The other Aleph pulled their arm back, ready to strike.

"Why was training canceled?" I asked, preventing

another fight.

"The Caretaker has not returned."

Simultaneously, they both started to tear up.

The Aleph holding the book responded between sobs, "She saved us from an owl. It was quite heroic. I would be gone if she didn't show up when she did."

The other Aleph visibly shivered in response.

"I hope she turns up soon," I said, placing a hand on my opal necklace. Its vibrations surged through my body, providing comfort.

I hugged them both and picked up the mug on the table.

"Where are you headed to?" They asked together.

"If training's canceled, what are we doing here?"

They looked shocked that I would even suggest there was a better place to be.

"Yeah, but that doesn't mean *we* can't practice," One suggested.

I couldn't bring myself to tell them no. Their small size and large pleading eyes made it hard to push back.

"Alright." I agreed.

They beamed with delight. With how poor Aleph was in class, it was surprising to see how much he knew. He told me random facts about planting berries and ancient martial arts poses. He shared his dream of visiting an ancient nearby temple, hoping to increase his aether sense. We shared a few laughs, and I had a good time.

I thanked him for the training session. The shadows

grew long and distorted as the sun dipped lower, a silent announcement of the approaching twilight. I returned to my hollow. The empty feeling from the previous night returned when I sat on my bed. Not wanting to burden Tia, I convinced myself to stay in my room until dark.

I placed a few berries in my mouth as I watched shadows crawl across the grove. Hours passed as I melted into my loneliness and became trapped within my thoughts.

"This is stupid!" I said, no longer willing to wait another minute. "She seemed to enjoy having me over yesterday. Why would she not want me there now?"

Nervously, I made my way down to Tia's hollow. Wriggles greeted me at the entrance.

"Hello, Wriggles," I said, voice noticeably shaking.

The beast still felt dangerous to me, but there was a desire to connect I didn't fully understand. I took a deep breath. Bravely extended my shaking hand to mimic the scratching technique I saw Tia apply last night. The squirrel leaned into the scratches. I smiled.

Satisfied, Wriggles playfully scampered up the tree and was gone.

"Tia!" I shouted, "Are you home?"

"Yes, dear." Her voice was caring and calm. "I was hoping you'd visit today. Come in, I'll make some tea."

I got comfortable in the same chair I sat in yesterday. The emptiness inside melted away. Tia's genuine interest in me made me feel special.

"How was training today, dear?" She asked.

"Um," I paused, unsure how to phrase my observation. "I suppose it was good. But a lot of dots are smaller than typical."

"That's strange. I haven't noticed." She responded while standing on her tiptoes to reach the top shelf.

I grabbed the tea leaves jar.

"Have you always been able to reach this jar?" I asked.

"Of course. I think the tree is getting larger."

I examined a mug. It looked normal in my hands. Next to Tia, it looked massive, though.

"I think you might be smaller," I said, holding out the mug.

She laughed.

"Do you always go around telling dots they are smaller than you?"

Wriggles climbed back into the hollow, cheeks full. A few seeds were visible from his slightly opened mouth. His pupils dilated, reflecting joy. Abruptly, nuts and seeds began spewing from his mouth.

"Ah!" I said, standing up.

Wriggles opened his mouth wider to allow more nuts to flow out. I soon stood on a small mound of nuts and seeds. Moist saliva glistened off the surface of the seeds in the candlelight.

Tia roared with laughter, rocking in place.

"Wriggles seems to like you!" she said.

As quickly as the animal came, it was gone up the tree

again.

A laugh bubbled up from deep inside me.

"He may bring you a few more before the night's end." She said. "It's his way of showing his appreciation."

I bent over to clean up.

"Leave it," Tia suggested. "Let's watch the evening and rest in the entryway."

I dropped the few nuts I picked up and wiped the slimy saliva onto my pants.

Tia provided me with an extra blanket. We sat on the floor and looked over the grove together.

Wriggles returned with a smaller load of nuts and added them to the previous pile. He then curled up next to us and provided back support.

His fur radiated heat. I leaned further into him and pulled the interwoven blanket higher.

"Thanks for letting me come over," I said as we watched the shadows overcome the forest.

"I'm glad you came back! You're always welcome here, dear." Tia stated.

As she blinked away sleep, the purple amethyst on her finger sent a radiant glow across her face.

"Is that a magic ring?" I asked.

"It was a gift to me from Al. He loves tea just about as much as I do. He promised this ring to me if I could find him more potent leaves."

The idea of more tea flavor made me contort my face into a grimace.

"He gave me this ring to protect me while searching the forest for unique tea leaves."

With a yawn, she stretched, reaching her arms above her head. The sun had disappeared. She removed her ring and laid it beside her.

"Unfortunately, I have to take the ring off at night. Otherwise, my finger goes numb and wakes me up. Stay as long as you'd like, dear."

She nestled herself into Wriggle's fur and closed her eyes. The thought of returning to my room was unbearable.

Just another few minutes, I told myself.

The quiet grove below added to the peaceful atmosphere of home, making me feel sleepy. I gave in quickly.

A horrendous and shrill noise filled the air. I jostled awake. Groggy and unclear, my brain tried to make sense of the situation. Tia was sleeping soundly next to me.

Could she not hear the noise? I wondered.

I watched her chest rise and fall in a gentle rhythm against the unsettling sounds that filled the air. Sounds she clearly couldn't hear.

Shapes and figures appeared in the darkest shadows, growing closer to us. Fear gripped me tightly. My eyes closed, and I nestled deeper into Wriggle's fur.

The passing wave left me feeling submerged in shadow. I lay rigid and trembling.

Moments passed in a terrifying nightmare blur.

I opened my eyes. The morning broke through the treetops, giving no hint of any noise.

I looked for Tia, but only saw her blanket and ring.

"I didn't know you were awake," I said.

There was no response. My heart beat faster. I stood up.

"Tia!" I shouted. "Where are you?"

I ran into the library.

Empty.

The kitchen was void.

The air felt icy. I returned to the blanket where Tia had rested a few moments ago. The ring lay motionless, offering no comfort or help.

Tia was missing. I was alone.

Tears flowed down my face as I stared at the blankets that had once held Tia. There was no sign of her exit. It appeared as if she had simply vanished, and the fabric collapsed.

In my despair, a powerful force welled up inside me. A deep burning desire to find her—avenge her if needed. The thought that she was still alive—a soft hope. It kept me crying.

I attempted to walk into the kitchen. My vision blurred, and I couldn't focus. It was hard to do anything other than cry. I somehow found the strength to prepare boiling water and placed some tea inside.

I curled up in a ball and sobbed. Thoughts flooded my mind. Every thought added to an existing tangled web. Each strand made it harder to see a way forward. A fresh pain jabbed at my throat—the deep, troubling sense of helplessness. There was nothing I could do to be with my friend right now. Breathing was getting harder.

As I tried to steady my breath, the sound of my ragged inhales and exhales filled the silence. The weight of helplessness settled in my chest, making it hard to breathe. I felt a fresh wave of pain, like a sharp stab in my throat, as I longed to see my friend again. The feeling of being unable to change the situation hung heavy in the air.

My body trembled with each gasp for air, my mind clouded with the overwhelming sense of loss. I could feel the wetness of my tears on my cheeks, the strands of hair sticking to my damp face. The pain pulsed through me, making it hard to focus on anything else.

I woke in Tia's hollow. The realization of Tia missing weighed heavily on my body, making it hard to move. However, having felt my emotions, I could think more clearly. The tangled web of thought was no longer there.

The noise. I thought. *It's what took her.*

Though unseen, the noise was a clue. I needed to find its source. A warmth swelled in my body, along with a determination to solve the mystery.

The light from the ambers flickered, the boiling pot of greenish water in front of me. A mug rested on the counter next to the pot. As I stared into the liquid, a new thought emerged.

Leave Countable Grove? You couldn't survive on your own before. Why would you leave the safety of this place?

The distant thought of my cold, helpless body lying on the forest floor flashed through my mind. My body shivered in response. I took a deep breath.

I am not the same person I was. I attempted to reassure myself. *They have taught me how to survive. I didn't have that knowledge before.*

"Ouch!" I shrieked, boiling tea burning my hand.

Wriggles showed up next to me, eyebrows pulling towards each other with a look of concern.

"It's ok," I reassured the animal. "Just me being stupid. Who will take care of you while I'm away?"

I stared into Wriggle's eyes. It ran towards Tia's bed. He smelled it and looked at me with concern.

"I don't know where she went," I said.

The creature chirped as if crying. Its breath was quick. I placed a hand on his side. His worried eyes rested on mine.

"I promise I will find her."

He sprang out of the tree and ran towards the top.

"Hopefully, squirrels can take care of themselves," I said.

I ran to the training grounds. One of Aleph's hopeless

plants remained on a table alongside his bag tucked inside a neighboring pot.

"Aleph! Are you here?" I shouted.

"Over here!" A muffled voice came from the other side of a plant.

"Are there still two of you?"

Jumping out of the bush, he stood in front of me, still the same size as yesterday.

"No, he was gone this morning. I'm glad, though. It was weird having another me around."

He laughed at the statement for so long that I wasn't sure what to do other than laugh with him. His energetic levels were contagious, and so was his enthusiasm for learning. I was glad to have him as a friend.

A sharp pitch echoed through Countable Grove, interrupting our laughter.

It's happening again. I thought to myself.

This time, I forced my eyes open. My hands covered my ears. Aleph remained calm, making no sign he could hear the noise.

"What's wrong?" he asked, a crease between his eyebrows.

There is no time to explain. I needed to see where the wave came from. My eyes darted around. A transparent dark wave flowed over us, followed by shadowy shapes. Dark green vines crawled and crept through the training grounds, circling Aleph's ankle and up his leg. They bound his body from head to toe and carried him off.

"No!" I shouted.

The vines were in full retreat, with Aleph bound tightly.

I chased them. Dots all around me were splitting into fractions—smaller, proportionally duplicated instances of themselves. Fifths became tenths. Thirds became sixths.

I sprinted after the vines until I reached the edge of Countable Grove. The shapes and figures moved too quickly for me to keep up.

I threw rocks and pebbles towards the vines, but I knew it was useless. Aleph was gone.

I stood inside the perimeter of safety until sunset. Fear lingered at the edges of my mind, whispering that I wasn't ready, that if I did try, I'd fail. But then I thought of Tia's smile, Aleph's excited ramblings, and Wriggles' gentle chirps.

They needed me. I couldn't let fear win.

"Give my friends back!" I yelled, sprinting into the field.

An owl cooed in the distance.

I glanced up and noticed the field of stars above. A gasp escaped my lungs. What a beautiful sight. The two moons were looming over the trees. I felt a deep connection to the twinkling lights in the distance. Something I could not see in Countable Grove.

An earth-shaking thud caused me to lose my footing.

I glanced towards the source. The ground shook from another impact, preventing me from standing—a massive figure hurtled toward me.

With great effort, I stood.

Each step the giant took shook the ground harder.

A burning sensation shot through my muscles as I fought to stay upright.

The figure towered above me. Large, sturdy fingers wrapped around my waist.

The wind swallowed my scream.

The beast lifted me into the air.

I gazed into its dreadful eyes.

"Zero!" the creature roared in delight.

They threw me in the air, and I slid into their satchel.

X

The Negative Caverns

Eulalia

My senses worked overtime to orient myself. The air was damp. I heard a rhythmic drip-drop of water nearby. The moisture of the rocky ground seeped through my pants and cloak. In step with my pulse, the aether within my opal necklace vibrated. The earthy scent in the air filled my nostrils.

I moved my hand in front of my face until I felt the warmth radiating from my palm. I saw nothing. Darkness consumed my vision. I felt around in my cloak. The Ishango knife was still in place. No berries, though. My stomach rumbled in response.

I leaned against a rocky surface behind me. Tingles flooded my sleepy leg as I readjusted. A shrill shock of cold followed every movement.

"Is anyone there?" I asked the void.

"Yep," a voice answered.

Relief washed over me.

"Ronan, is that you?"

"It is. I'm glad to hear your voice. I've been awake for sometime now and thought I was alone."

My body shivered. I tried to not move to preserve the warmth soaking into the cold stone.

"Have you ever seen darkness like this?" Ronan asked between gritted, chattering teeth.

A third voice jumped in and jokingly stated, "You know, it's strange to talk about darkness as if you could see it."

"Ellis!" We said.

Delight filled my body. *All alive—all together.*

My damp pants clung to the side of my legs. With every shiver, a sharp wave of icy coldness rippled through my body. Prompting more involuntary movement.

"Only Ellis would think about something as philosophical as sight in a dire situation." Ronan said sarcastically, with a friendly chuckle.

"Might as well keep myself entertained." Ellis responded.

The three of us shared stuttered laughter between chills.

A flicker of dim light glistened off the damp floor.

Ellis and Ronan's bodies were now noticeable as flat, dark outlines.

A fourth figure lay motionless on the other side of Ellis. The subtle light grew to a medium greenish glow, revealing the cavern walls that resembled a dungeon. The room was completely sealed except for a crevasse across from me.

Growing brighter, I caught a clear view of the dot nearest the exit. They, like Opal, had cuts and bruises across their whole body. Their torn clothes revealed their bone structure.

"We need to find a way out of here," Ronan whispered between shivers.

The light source, a dot from the negative domain, stepped into the dungeon with us. Like a greenish-blue campfire ember, the dot was translucent. The dots light fully illuminated the room in a teal color. I've never seen them in the dark like this. When they are in the sunlight, they appear almost see-through.

Ellis stared in shock.

Ronan shielded his eyes from the brightness.

The dot pulled a few berries from a bag and offered them to us.

Cautiously, I approached.

"Are we prisoners?" Ellis asked.

"Can prisoners choose to leave when they want?" The glowing dot responded.

"How can we leave?" Ronan jabbed. "We can't see anything."

"Yes, but the choice remains yours."

I grabbed the berries and started making preparations for a duplication spell.

"Magic taints these berries," the dot cautioned. "They are already four generations past their effectiveness. Casting your spell will only make things worse."

It was disappointing to hear.

The three of us looked at the unknown dot curled into a motionless ball. We shared glances of concern.

"We will need to care for them," I said.

"I wouldn't expect anything less," Ellis stated admiringly.

We split the berries between the four of us. They tasted like berry-flavored sand.

"It would have almost been better to have nothing." Ronan whined.

I grabbed the leftover berries and compressed them, making a pasty juice. Ellis held the dot's mouth open as I placed the paste in their mouth.

They swallowed—a good sign.

The transparent dot said nothing and left.

With their back turned, I gestured to follow them out.

Ronan, the strongest, threw the unconscious dot over his shoulder.

We followed behind, attempting to remain out of sight. We walked down a long, dark corridor. Cracks in the walls showed empty cells, reminding us of the one we escaped. They were all vacant.

"I know you're following me," the dot from the negative domain said. "There is no need to sneak around. I am a guide of The Negative Caverns. Most call me Lyra. My job is to help dots like yourself find their way in the dark."

"What about this fellow?" I demanded, stepping out from behind a rock. "They do not seem to get the help they need!"

"We have limited supplies for natural born dots. Dots of this domain receive energy from the naturally growing mushrooms in the cave. We only need berries from the surface when dots like you get lost down here. Fortunately, it happens less often than you might think."

"Why not give them mushrooms?" Ronan asked.

"Any natural-born dot that eats will forever take on this form. We consider this choice to be the most sacred. To force a dot to transition is the worst transgression possible. The dot cannot agree, so we try not to intervene."

"Can *we* have some mushrooms?" Ellis asked.

Lyra pulled a few from her bag. Three of them rested in her palm. One red and the other two blue. Their light mixed with the teal glow of her skin.

"If you want," she said, holding them out to Ellis. "We do not know of a way to reverse it. Most dots that convert stay here. It's too dangerous for us on the surface."

I could see the weight of hunger on Ronan's and Ellis's faces. We couldn't risk not being able to leave. There is too much at stake.

"No thanks," I said dismissively. "Let's stay focused on

getting out of here."

Lyra placed the mushrooms in her bag again and pressed on.

"It appears we are in the negative domain." I whispered to Ellis.

"Yes, but how did we get here?" She asked.

"The cloaked figure must have brought us here. We have to assume this dot is working with them. We cannot trust them."

We followed Lyra down a long hallway that opened to an enormous cavern.

The colorful glow of mushrooms lined the walls. Before us was a massive staircase to take us deeper. Darkness splashed across the rocky landscape between clumps of mushrooms.

In other conditions, this view would be spectacular.

Though unchained, I felt like a prisoner. Not being able to see provided a severe detriment to the possibility of survival.

"Watch your step. The ground can be slippery." Lyra said and proceeded down the staircase.

Hours passed as we walked.

"Do you mind if we rest for a minute?" I asked.

We all stopped and found the nearest rock to sit on—the sound of a river was close.

Lyra waited patiently, showing no signs of fatigue.

"Do you have any more berries?" Ronan asked, sounding agitated.

"Unfortunately, no," she responded.

I glanced just beyond the Lyra's glow to see a rickety old bridge.

"I hope we do not intend to cross that bridge," I said with a head nod towards it. "It will not hold more than one or two of us simultaneously."

"Alas," Lyra said playfully. "We will cross that bridge today. There are seven bridges spaced through the caverns. Luckily for you, we only need to cross this one. Consider it a welcoming challenge to show the leaders you are worthy of a visit."

"Challenge?" Ellis shouted, fuming with rage. "Why would you have such a thing?"

"What if we fail?" I asked, still trying to catch my breath.

"If you fail, the cavern will claim you forever." Lyra showed no emotion in her words. "Upon your success, you will receive a proper welcome, including food, shelter, and a way back to the surface."

We threw up our arms in frustration. Our grunts and groans reverberated into the darkness.

Ellis screamed in rage. The shrill noise echoed around the cavern walls, dimming into nothing.

Ronan carefully placed the unconscious dot on the ground. Then he charged at Lyra, arms outstretched to grab her by the shirt, but passed through as if she wasn't there.

"If you prefer, I could leave you here," Lyra said calmly.

I placed a hand on Ronan's shoulder.

"Get hold of yourself," I said.

They both took calming breaths.

The dot we had carried from the dungeon moved.

"Why is it so loud?" They groaned.

"Sorry about the noise," Ellis said between gritted teeth.

"Can you walk now?" Ronan asked aggressively.

He seemed surprised by his own words.

"Sorry, I didn't mean to sound so aggressive," Ronan responded. "I'm just hungry, tired, and need some rest."

"I could try."

Their tattered clothing swayed as they struggled to keep their balance. Ronan helped provide sturdy support as they wobbled in place. Their legs gave way as Ronan lowered them gently to the ground.

In labored breaths, they said, "Looks like I'm still too weak."

Ellis gave an encouraging smile.

"I can carry you for a while." She said.

"I appreciate it." He responded. "I have been here for weeks. In and out of consciousness. I thought for sure I was going to die."

"And who, exactly, are you?" Ronan asked, trying to force a kinder tone and failing.

"My name is Oizys," He responded casually. "A name given to me because I seem to always find myself in trouble."

"What kind of trouble?" I asked, aware that we didn't need more problems.

"My friends and I were exploring the caverns when a rock gave out, and I fell in. Another time, we encountered a bear which attacked me but left others unharmed. Just mysterious misfortunes."

The dot looked like me and was about my size. He must be from the natural domain. He spoke of a life outside of Origin Tower I didn't know existed.

"Do you know Arithma?" I asked and studied his face.

"No, can't say that name rings any bells. My friends live on the other side of the inverse mountains."

His face didn't reveal any signs of lies. Arithma never mentioned dots living there. I thought all dots were born in the birthplace at Origin Tower's base. If what Oizys said was true, there are dots elsewhere, too.

Ellis helped carry him to the bottom of the stairs.

The stairs gave way to an open platform of dirt. The edges had a colorful glow from the fungus spread over the rocks. A prominent stone rested in the center.

The luminescent glow revealed engravings on the large stone.

A bridge between worlds, five hundred paces long,
Tread lightly, lest your steps go wrong.
No more than two may pass at a time,
Seventeen minutes is the span of the climb,
Or the leaders' eyes will see you no more,
Lost to the shadows beyond their door.
Let hunger drive you, the quest will begin —

For a feast awaits in the leader's den.

Before I had a chance to fully process, Lyra started across the bridge.

"Wait!" I yelled.

"I will see you on the other side." She yelled back. "Use the torch to get across when you're ready. I will only add time and make it impossible for you to cross."

A torch flared to life on the engraved boulder's surface.

Ellis said, "Oizys, I think you must cross on your own. Are you up for the challenge?"

They looked around for a few moments and then tried to stand up. Their knees wobbled, and Ronan and I helped them get their balance.

We walked to the edge to better understand the challenge ahead. The bridge spanned the river nearly fifty yards below. Slats were missing, and others displayed visible fragility as if any weight would cause them to break free.

"There is no way I'm crossing that bridge," Oizys said.

Ronan responded seriously, "I wish we had another choice."

I re-read the poem. Seventeen minutes in total. Only two at a time. I ran calculations in my head.

Ronan could cross in a minute.

Ellis, I'm guessing, would take twice as long.

Oizys about 10 minutes.

I was slower than Ellis and Ronan. I guessed about 8 minutes for me.

"We only have one torch?" Ellis complained. "We can't risk losing the torch in the river below."

Ronan jumped in, "You're right. If I go with Ellis. We will send one of us back. Then take each of you."

Oizys shook his head, unwilling to move forward. He said, "You guys can't seriously be considering crossing that bridge. Is it the only way out?"

Wishing there was another response, I said, "Lyra is our guide, and without her, I don't see another way out."

He agreed to work on getting across. My mind returned to the problem ahead of us.

If Ronan goes with Ellis, they will take two minutes to cross and then one minute back—three minutes. Then I go with Ronan. That would be 9 minutes for me and one minute back—thirteen minutes. Another ten for Ronan and Oizys. Twenty-three minutes in total.

That's too slow.

Ronan and Ellis stood in front of the bridge, prepping to cross.

"No! Wait!" I shouted. "It's too slow. We need a faster solution."

They turned and looked at me.

"I just need a moment to think it out." I pleaded. "Let's slow down and see if anything new comes up."

They stepped away from the bridge. Ronan held the torch.

The four of us agreed that if we couldn't find any other solution, we would just try to cross the bridge as quickly as possible, hoping to make it in time. Hunger and fatigue made it hard to focus.

The group's frustration mounted as we repeatedly hit dead ends.

"Here's a crazy idea." Ellis joked. "What if we send the two slowest dots together?"

We all laughed.

The joyous moment was a welcomed distraction. For my sanity, I considered the possibility of sending the two slowest together—*if we send the two at the beginning, that is an obvious issue.*

My heart beat faster.

"Ellis, you're a genius!" I shouted. My voice faded quickly into the roar of the water below. Her cheeks turned red with embarrassment. She looked away.

"C'mon Eulalia! Don't be ridiculous." She said.

She usually called me The Caretaker. It was strange to hear her say my actual name.

I stood up again to get a good look at the bridge.

"No, seriously. We won't send us both at first. But what if we send us second?" I asked, turning back to the group. "If Ellis and Ronan go across first. Ronan can then bring the torch back to us. Oizys and I can cross together, and Ellis can bring the torch back to Ronan."

Ellis caught on. She stood up, beaming.

"Ronan and I will take 3 minutes to cross." She said. "We get a two-for-one with you and Oizys! Making it 15 minutes when I make it back to bring Ronan across, too."

A sensible plan in place, we found ourselves at the bridge's edge. The bridge seemed even more rickety than before.

"Alright. Whenever you're ready." Oizys said, looking nervous. "I know there is a time restraint, but take your time crossing."

Ronan and Ellis made crossing the bridge look easy. They pointed out faulty slots to make it simpler for Oizys and me to cross. Within moments, Ronan was back on our side with the torch.

Oizys and I acknowledged each other and started across the bridge. It swayed and bounced under our weight. The ropes on either side were old and wet. The moss made it more slippery and difficult to get a good grip.

"Watch that step!" Ronan shouted behind us.

"Try to walk on the rope for the next three slats," He coached.

We were nearing the end of the bridge. My heart was feeling lighter—just five more.

A crack rang out.

Oizys fell through and slammed against the side of the bridge, causing the whole thing to rock violently.

I clutched the torch as I turned quickly to help.

His arms wrapped around the slippery rope.

I carefully lowered myself to offer my hand to help.

"We've made it this far," I said encouragingly. "Might as well finish it out!"

He reached for my hand, and I pulled him onto the

bridge. My heart pounded in my chest, taking in the reality that we had cheated death. I could see the relief in Ellis's eye as we made it fully across.

I gave the torch to Ellis and fell to the floor, trying to catch my breath. I didn't want to do that again. Ellis and Ronan came back before I could sit up completely. They helped me stand up.

"Well done!" Lyra said. Standing near us. "You are the first dots to have completed this challenge."

I cringed at the words. The broken slats were the tombstones of previous attempts. A grim reminder of those who didn't make it though.

We walked around a narrow pathway as the roar of the river faded into the distance. Wedged between two cliffs was a village of large mushrooms. Four or five different colors glowed brightly. Transparent dots were playing cheerfully. A bio-luminescent fungi path led from the cliff's base into the center of the village. We no longer needed Lyra to guide us, and we could easily find our way along the paths scattered throughout the village.

We took in the breathtaking view.

"You promised us a feast," Ronan said to Lyra.

"Yes, of course. Follow me." She responded with a bow.

There was still a nagging question in the back of my mind. *Who is the hooded figure? Why did they bring us here?*

XI

Mortimer and Plex

Opal

I clawed at the silky fabric as I slid further into the gnasher's satchel. Unable to hold on, I closed my eyes and braced for impact with the bottom of the bag. The soft fabric broke my fall, leaving me unharmed.

My nose picked up on a wonderful aroma of roses with a hint of berry. A dim teal glow provided enough light to see the gentle folds of the fabric.

This isn't so bad. I tried to reassure myself. The gentle sway of the beast combined with the smell provided a false sense of comfort. I sensed danger, yet I couldn't reconcile the threat with such a delightful experience. I allowed my body to relax into the cradle.

The hideous orange and green eyes of the beast flashed in my mind.

My breathing sped up as realization settled in.

I'm trapped. I thought. *Before I even got started. How can I be so stupid? Tia, Aleph. Both of them are gone.* I felt a heavy weight on my chest. My vision blurred. To fight back tears, I made fists—anger took hold of me. A burning sensation replaced the weight on my sternum. My back tightened.

"Let me go!" I screamed.

A voice from somewhere in the satchel responded, "He finally caught another one?"

I took a shuttering breath in as I sat up. The beast's swaying motion made it difficult for me to keep my balance. The source of the light was a semi-transparent dot.

He masterfully skipped to me, every step perfectly timed to the rocking motion.

"I can't believe it!" He shouted. "Are you real?"

His eyes glowed brighter than the rest of their body. His clothes softened the light emitted from his skin.

"Oh... my god! I am talking with another dot," he said, bubbling excitedly. "I've been in this bag for who knows how long."

The transparent dot shouted towards the opening. "You finally caught one! I can't believe it! I'm no longer alone!"

Were they celebrating? Rage surged through my body. I

attempted to stand up but rolled forward.

"You will get used to the movements," the dot reassured me as they danced over to my new location.

"Why are you so excited about my capture?" I demanded with a scowl.

They looked shocked but pleased.

"You really *are* real." They stared at me in disbelief.

"Of course I'm real. Who are you?" I asked.

They bounced around, giggling with anticipation.

"I'm so sorry," he said, snickering—eliminating any sincerity in his apology. "Your capture means..." He put his palm to his head as if hitting it would help him find the right words to say. "So much to me. Wait! That didn't sound right."

"Excuse me?" I said in disbelief.

His hands tried to soothe the air.

"No, sorry. That was stupid. I'm no longer isolated. I've been here so long I've lost track of the time. Occasionally, Plex will show up for a bit but comes and goes randomly."

"Wait, what?" I interrupted, "Someone else is in here, too? Where are they?"

"Well... No... I am Mortimer. Plex is my brother. We're from the complex domain."

I gave a confused look.

"You know..." He said expectantly. "We drank from the complex sea, and now there are two of me. Normally, he is only in my head. But lately he just shows up unannounced and takes over. It's rather annoying."

"Your brother can leave the satchel?" I asked excitedly.

"Well, not exactly. My brother is like a second part of me. I've only seen him within the last few weeks. Before, he just existed in my head."

Mortimer stared at me, looking for something—a confirmation, maybe. The deep stare was unsettling. The bag swished in a new direction, causing him to fall on his butt.

This guy is a lunatic. I thought.

"Your brother, is," I paused, trying to reconcile his words, "part of you?" I asked.

"Well, it's hard to explain," He said in frustration. "I wish I had a way to show you. The first time it happened, I was in The Negative Caverns, relaxing in a hammock. Out of nowhere, *Bam!* I had a second head. Then Bam! I'm in a gnasher bag."

"Oh," I responded, "you mean Plex is *literally* a part of you."

"Yeah. When Plex is in control, I can only catch glimpses of what he's doing. It feels a bit like a nightmare."

"Right," I responded. Uneasy about my strange satchel mate, I attempted to change the subject. "Do you know how to get out of here?"

"Hey, don't go!" Mortimer said, looking concerned. "I need someone to talk to. I haven't spoken out loud for at least a few days now."

Frustrated, I crawled toward the nearest seam, hoping to find a loose stitch to crawl through.

"Gnashers sure know how to make nice cloth." He said, dancing around me as I began inspecting the edge of the bag. "I've been admiring the softness for quite some time. I've got some of the best sleep of my life in here."

He continued to talk while I searched.

"Before getting caught, Plex and I became friends, actually. Initially, I hated having Plex with me. Do you know how frustrating it is to have someone controlling half of your body? He would interrupt everything all the time. He interjects his own opinions and feelings into my thoughts. It's like a constant battle in my head."

Mortimer moved in front of me to ensure he could see my face as he continued, "I would say, 'I like mushrooms'. And he would respond, 'I like berries'. And then he would make me eat a berry!"

The disdain on his face was palpable.

"Do you have any idea what berries taste like? The worst things on the island! Why would anyone eat them?"

He paused, waiting for me to agree, then continued before I could argue.

"It was rather depressing, but he helped me see I didn't fit in with the other dots. I never thought he would leave The Negative Caverns, though."

"Shut up and help me find a way out of here!" I shouted.

He stopped talking immediately. His shoulders and face dropped.

"You don't want to be with me. I understand."

His reaction made it obvious he, in fact, did not

understand. Head hung low, he walked to the edge of the bag and pretended to check for seam flaws.

Guilt slammed into my gut.

Was I feeling sorry for him? I wondered. *What if he's telling the truth?*

No, He's crazy. The sooner I get out of here, the better.

As I continued to feel for loose threads on the seam, I noticed his energy had returned. He playfully mimicked my search as he moved in my direction, pretending he couldn't see me. We were moving closer together until his hand passed through my arm.

"Oops." He giggled, attempting to look innocent. "I didn't see you there."

He rolled on the floor, laughing.

I sat back and observed, unsure of what to do.

His playfulness was genuine, and his laughter was contagious.

I smiled and crawled closer to him. I reached out, and my hand passed through his arm.

"Oh, no!" I joked. "I didn't see you there."

His laugh roared louder.

For a moment, I forgot we were captives. I forgot he was crazy. We continued to joke as our limbs passed through each other.

My opal necklace fell out of my shirt and dangled from my neck. When my hand passed through his foot, it glowed orange.

A gasp escaped my lips.

"You didn't tell me you were a witch," Mortimer said, sitting up and looking at the gemstone.

"Oh, I'm not. This was a gift from a friend," I responded, also sitting up.

I held up my hand, and Mortimer reached out his. Our fingers brushed lightly, the necklace glowed dimly in response.

"What does it mean?" I asked.

Mortimer and I stared at the mesmerizing colors shining off the reflective surface.

He said, "A wizard in The Negative Caverns once told me gemstones glow when magic is performed. She told me about a magic that allowed dots to combine into one form. She said aether was all around us and unified all dots, including the plants and animals. It seemed like nothing more than a good story, but seeing the glow of this necklace, I wonder if it's possible."

"Like two dots existing in one form?" I asked.

He shrugged. "Sure, I guess. I never really understood it."

I pulled my hand back and stared into the smooth surface of the opal gem.

"No offense, but you won't find anything wrong with the edge of the bag. I've looked for a weak point for days. Gnashers sure make sturdy material."

I felt like I was observing an utterly different dot. He seemed normal. I felt myself giving in to his ridiculous story. Plex, his brother, who I assumed was imaginary, was a part of him but only showed up occasionally and recently...

The noise!

"Mortimer!" I said, standing up, feeling a surge of excitement.

He turned to face me. The gnasher swayed, and I fell towards him.

The necklace glowed brightly as our bodies merged into one. We fell through the fabric and slammed hard into the ground. Both of us were moving as one.

We stood up, and I pulled our hands in front of us. A faint misty outline showed the borders of our arms—as if we no longer existed. The ground shook as the gnasher continued to walk away from us, making no sign they noticed our absence.

I stared, puzzled.

The necklace glowed brightly in the dark evening. I pulled it off, and we separated.

Mortimer tapped his sides as if checking that he was still there.

"How?" Mortimer placed his palm on his forehead, looking for words.

"Where did," he continued to struggle. "You are a witch! That was incredible!"

My face broke into a smile. It was good to be out.

"I didn't mean to do the spell," I responded.

"Well, you did."

He took a deep breath of the night air.

A cool breeze made me wish I had Tia's blanket. Mortimer's glow provided light so we could have a dim view of our surroundings. We stood in an open field with boulders sporadically spaced all around.

In the distance, beyond the boulders and forest, was an odd structure. A dark outline of a massive tower reached high above the treetops.

"I never thought I'd get out of there. Thank you!" Mortimer said, with tears.

"Wait, is that rock moving?" I asked, staring at a boulder.

Mortimer looked to where I gestured.

"That's not a rock. That's a gnasher." He said it matter of factually.

The scene took on a new perspective. If every boulder was a gnasher, hundreds of them surrounded us. Quietly, we approached the nearest entrance to the woods.

A wave erupted from the dark tower beyond the forest. An ear-splitting shrill noise filled the air as the wave flowed towards us. My hands covered my ears, and I shrieked with pain.

Mortimer looked at me with concern.

"Is everything ok?" he asked as if he couldn't hear or see the wave.

The wave flowed over the gnasher camp. I tried to keep my groans quiet.

Mortimer's body bubbled, and then a second head and neck grew between his shoulders. His glow went out, sending us further into the darkness of night.

The shrill noise dissipated.

"You got us out of the bag!" The head that could only be, Plex said in surprise and rather loudly.

"Shut up! Would you?" Mortimer said, placing a hand over his mouth and nodding towards the gnashers.

Once Plex was quiet, Mortimer whispered, "I didn't do it. She did."

He nodded in my direction.

"She's a witch!"

It was surreal to see them both together. Two heads attached to one body.

Mortimer wasn't lying.

Plex looked at me.

"She doesn't seem like much of a witch," He said. "Are you sure she is how we got out of there?"

I was speechless. "He–You?" is all I could muster.

Plex grinned.

"Did she think you were crazy?" He asked Mortimer with a laugh.

"Of course she did!" Mortimer responded. "Now that you're here! I've got a few words for you."

Mortimer smacked Plex in the back of the head.

"Hey!" he retorted, struggling to keep quiet. "What was that for?"

"How did you get us trapped in the first place? You idiot!" Mortimer responded in irritation. "I was alone for days. Next time, have the decency to stick around for a bit!"

"Hey! You do not know what we went through." He argued. "That gnasher was the lesser of two evils. I saved our lives getting in there."

Mortimer's mouth dropped. His bewilderment was quickly replaced with anger. He pulled his hand back, ready to strike again.

"Quiet. Both of you!" I interrupted. "We're still surrounded by gnashers. This is hardly the place to argue."

Plex stuck his tongue out, and Mortimer hissed in response.

I lead them towards the woods. Even in the dark, large black silhouettes made it easy to tell where the gnashers rested. Not all of them were asleep, though. A few smaller ones were playing tug of war with their satchels.

My thoughts immediately went to the dots trapped inside and my mouth filled with a sour taste.

Another wave burst from the tower.

The noise was unbearable.

I fell to the ground, covering my ears. Unable to keep myself quiet, I screamed in pain. Plex moved to cover my mouth.

Mortimer's head dissolved into Plex's body, leaving Plex behind.

Despite Plex's efforts to silence me, it was too late.

The ground rumbled. Ten tree-sized gnashers stood ominously over us.

"This way! Quickly!" Plex pulled me by the arm.

He ditched me and ran the opposite way from the

safety of the trees.

Why would he go that way? He seemed so confident, though. Realizing I may live to regret the decision, I followed.

Luckily, my legs cooperated this time. With each step the gnashers took, I anticipated the shock waves. I quickly caught up to him.

"I hope you know what you're doing," I said disapprovingly.

We wove through the chaos, dodging massive arms that swung down like tree trunks.

Slam!

Two gnashers collided with a bone-rattling impact, their arms creating a brief archway of safety. Heart pounding, I sprinted beneath them, barely squeezing through before they crashed together again.

On the other side, I ran face-first into Plex, who had halted. I stumbled back, clutching my nose, blinking away the tears in my eyes. Then I saw it.

A sea of gnashers stretched out before us—hundreds, maybe thousands—loomed like jagged boulders under the dim glow of moonlight. Moving forward was impossible. The sheer number made me feel hopeless.

Panic clawed at my chest as I scanned the area. A tiny bush caught the corner of my eye. It wasn't much, but it was something. I darted toward it.

"No!" Plex's voice cut through the chaos as he grabbed my arm.

Before I could argue, a colossal fist crashed where I'd

been standing.

Plex pulled me sharply to the side, and we tumbled into a narrow gap between two gnashers. He moved with purpose, his confidence cutting through my confusion.

I followed him, ducking and weaving as shadows of giant limbs swept overhead.

Suddenly, Plex dropped to his knees, brushing aside loose soil and revealing a narrow hole in the ground.

"Quickly!" he hissed with urgency. "They can't follow us in here!"

The gnashers' thunderous footsteps grew closer.

I jumped feet first into the hole. The earth trembled above me as the gnashers collided with another deafening crack. Dust and pebbles rained as the sound wave of two large, colliding masses pulsated through the narrow tunnel.

We caught our breath at the bottom of the hole.

"I'm glad you knew about this place," I panted.

"The Negative Caverns are massive," he said between breaths. "You can typically find a slot here or there where you can jump in for a bit."

The darkness swallowed everything, leaving the cavern's size and shape a mystery. I leaned against the cool earth, catching my breath. My fate was clear. The thought of stumbling mindlessly back into the gnashers' territory made my stomach twist. Without light, without direction, survival felt like a fleeting illusion.

XII

Dux, Phi, and Quint

Eulalia

Lyra waved her hand. A few dots standing guard open a pair of heavy doors. A pleasant aroma filled the air. Our steps echoed around the dining hall. A colorful display of mushrooms lined the room's base, highlighting the central table. Which overflowed with berries, mushrooms, and mysterious dishes.

"Your reward," Lyra encouraged. "Eat as much as you'd like."

A vibrant array of colors filled the spacious room as negative dots filled the glass balcony above. Each vibrant glow cheered as they made their way to the center.

Oizys was the first to the table.

I cast my skepticism aside and joined him. Each bite filled my body with renewed energy.

"What is this thing?" Ronan asked, holding up a triangular-shaped pale item.

"We call it a pastry," Lyra responded.

Ronan gave a satisfied groan with jam smeared across his face.

Unable to shove another bite into my mouth, I moved to sit in one of the twenty chairs lining the table.

"These mushrooms are safe for naturals to eat," Lyra said. "They will not make you transparent."

I picked one up. The mushrooms didn't have the same glow as the other mushrooms in the caverns. They smelled earthy and felt fleshy. I shrugged and took a bite. I immediately spit it back out. "I'll stick with the berries."

They felt dirty in my mouth. Like I had just licked a rock or patch of moss.

Ellis loved them.

Relieved to have a full belly, I felt a sudden chill drawing my attention to the head of the table.

A hooded figure dressed similarly to the one that captured us rested in the center seat with two large chairs on either side.

I dropped the remaining food I was holding.

Ronan gave a curious look and followed my gaze. Once he saw the figure, he nearly choked on the pastry stuffed in his mouth.

Ellis placed the mushrooms down on the table and gave me a look of concern.

Oizys made no sign he recognized the heightened tension.

"Don't stop on my account," the figure said in a deep, authoritative voice.

Another cloaked dot stepped into the room, followed shortly by a third. Their heavy cloaks concealed their bodies entirely. They sat in the flanking chairs at the head of the table.

"Too scared to reveal yourself?" Ronan challenged, fists tense and ready for a fight.

The chatter from negative dots on the balcony fell silent.

Oizys grabbed a few more pastries and took a few steps back.

"Your caution is reasonable," said the middle hood.

He placed his hands on the edge of the hood's fabric.

The cloaked dot on the right stepped forward and snatched their hand. "We can't trust them." They whispered, a hint of fear in their voice.

"Nonsense!" they responded and pulled off their hood.

My heart dropped. The vines flowed up their neck and surrounded their head, restricting them to their rational form. Like Thrall, The Pythagorean Beans plagued this dot. They were a bound irrational.

He must be one of Arithma's guards. I thought. *After all this time, she finally found me.*

"My name is Dux, and I don't expect you to trust us," He said.

Ellis, full of rage, slammed a fist on the table.

"You got that right!" she yelled.

"May I be the first to apologize for the conditions in which we brought you here. We needed to know your intentions."

Ronan's ears turned red as he paced the floor.

"You could have just talked to us!" Ronan shouted in a booming voice that echoed against the walls.

Dux held a hand towards the figure on his right.

"Phi is the one that captured you." He said.

My mouth fell open. The tension in my back eased.

What kind of game is this? I thought. *They starved us, and put us through a ridiculous challenge. Fed us, and now they are telling us who captured us.*

Hands shaking, Phi removed her hood. She was lovely, with flawless skin. Unlike other bound irrationals, the vines looked natural on her. They provided highlights to her already beautiful form.

Dux said, "Phi was out scouting and noticed a group of naturals casting a concerning spell. Placing you at the scene of the crime twice is what prompted her to act quickly."

All at once, the realization settled in. They thought we were the enemy. I felt a flurry of emotions. Frustration because I could relate to their quick action. I felt anger because of their mistreatment. Sadness because Arithma is making progress and we have less time.

"So, this is all just a misunderstanding?" I asked.

"I think so," Dux responded. "You have shown tremendous character today. You cared for a dot that was not your responsibility. It is this act that lowered my suspicion of you. Our methods may be unorthodox, but they are effective. You have gained my trust."

His words cut deep.

"Unorthodox?" I said. "You nearly killed us!"

Before he could respond, Ellis asked, "How do we know we can trust anything you say? You lied to us about having more berries. What else are you lying about?"

Ronan, still pacing, nodded in agreement.

Dux appeared to have a loss for words.

"We need your help." The third hooded figure said, still in shadows of their cloak.

I stared in disbelief, shaking my head.

Ronan paced faster.

Ellis and I shared a glance.

"No!" Ronan responded. "I have an effective method of my own. Never help your captors!"

A conflict rose inside me. Dots willing to use such questionable methods seemed no better than Arithma. However, my experience with Arithma proved there were legitimate times to weed out allies from foes. I didn't know enough about these dots. Initially, I thought they were Arithma's guards, but she would never favor caring for the dot. She would consider the behavior a sign of weakness.

She would have punished us.

Ellis seemed trapped in a similar moral quandary. Any chance of trusting them required more. I took a calming breath.

"Dux, is your name?" I asked.

"It is." He responded with a bow.

"It is my understanding that any dot plagued by The Pythagorean Curse is under the service of the dot that cursed them. Who cursed you?"

His eyes grew heavy as he collapsed in his chair.

Phi placed an encouraging hand on his shoulder.

"I was the first dot to receive the curse. The dot who cursed me was the creator of the brotherhood."

"That would make you thousands of years old," I responded in disbelief.

"Yes, the dot that cursed me passed away long ago. The vines prevent me from using my full powers."

"What about Phi?" I demanded.

"I don't need to tell you!" She scoffed.

"Then we won't help you!" I responded.

"What do you hope to gain from these probing questions?" Dux asked.

"Arithma is the name of the dot attempting to wake The Cursed Dragon. If she is the one that cursed you, there is no way for you to gain our trust."

Sobs came from the third cloaked figure.

"It's not my story to tell," Phi said, moving over to the third figure.

"Quint," she said, "I will support you either way."

With a trembling hand, Quint removed her hood.

The vines wrapped around her face like thin green serpents, their restriction contrasted with the straight, perpendicular scars etched across her skin. It appeared as though something had tried to cut the vines. She stared at the corner of the table—avoiding any eye contact.

"My name is Quint," she blurted out. Her cheeks flushed with embarrassment. "Arithma was not the one to curse me or Phi. It was my fault."

Chatter erupted from the glowing dots above.

"Perhaps it's best we continue in private," Dux said with a wave of his hand.

The dots on the balcony left the room. The mushrooms on the walls became the only source of light. Oizys snacked on nearby pastries again.

Dux placed an encouraging hand on Quint's back.

"Naturals have been afraid of irrationals for as long as I can remember. Perhaps it's because of the unpredictable patterns or our heightened connection to aether. I hoped we could all live together. I heard about the beans and found them fascinating. My sister Phi cautioned me against seeking them out. I was too young and foolish to listen."

She let out a deep sigh. A chill filled the air.

Ellis shifted uncomfortably in her chair, her eyes flicking between Dux and Phi.

"Over a campfire, I told the story of the brotherhood to

my friends. I told them that the brotherhood believed that everything could be rational. We tried to imagine what it would be like to no longer have shifting heads. To appear natural-born."

Her voice quivered.

"I thought if we could be like them, they might not be as afraid of us. My friends and I went on an expedition to uncover the secrets of the brotherhood. We never made it inside."

Phi was fidgeting with the edge of her robes, anger in her eyes.

Ronan crossed his arms, brows furrowed.

"A terrible beast guards the Pythagorean Crypt. The beast lives inside the moat of cursed beans. Contained in the beans are the most horrible creatures to have ever lived. Each one pleads to be released. Begging anyone who comes close to free them. We never stood a chance."

Concern for Quint replaced the pained anger on Phi's face.

"At first, it felt like an adventure. We were going to free ourselves and become natural-born. As we got closer, the spite and hate of the beans overcame us. We found ourselves encircled by a serpent—poised to strike—fangs as long as my body. One by one, the beast took my friends."

"How did you make it out?" Ellis asked.

"Phi rescued me. In her normal form, she has the power to control nature. I was too stunned to fight back. She jumped between me and the beast. She skillfully pushed the snake back into the moat. Thinking we were safe, my foot clumsily fell in. Vines sprang up and pulled me under. She grabbed my arm and used what little magic she had left to grow a bush to push me out. We were safe, but the effects of the magic were too strong. The vines grew across my body and onto hers. We remained cursed."

The room was silent. Ronan gave an empathetic look towards Quint.

I believed their story.

"What is it you need help with?" I asked.

There was a recognizable shift in the room. Dux waved his hand and glowing dots filled in again.

"We need your help to uncover the secrets of The Pythagorean Brotherhood."

Ellis's jaw dropped.

"You're asking *us* to go inside the Crypt? After that story?" She asked.

"It's our only chance of learning anything to break the curse," Phi said. "We know there are secrets hidden inside. Irrationals are unwelcome. The Pythagoreans were natural born, which means you might get in."

Though hesitant, I wanted to help.

"We are on a quest to defend the land against The Cursed Dragon." I said. "We need to unite the five domains. It's our only chance to fight against what's coming. I will help you if you join us on our quest."

There was an exchange of looks.

"Done!" Dux said in excitement.

"Allow me a chance to confirm with my friends. I don't want to commit them to something they are unwilling to do."

Ronan, Ellis, and I huddled around the furthest corner. Both of them looked concerned. I placed my hand on the table. Cold fibers offered reassurance. The Crypt loomed like a shadow in my mind, coiled with secrets and danger. But if we walked away now... what hope would there be for stopping Arithma? I swallowed hard.

"We can arrange for your release and you can return to Countable Grove."

Ronan and Ellis shared a glance.

"Are you crazy?" Ronan said.

"You think we will just leave you to do this alone? No way!" Ellis interrupted.

I smiled, feeling relieved.

"Very well then." I turned to face the three leaders. "We will help you!"

A roar filled the space. The transparent dots on the top floor screamed in approval.

Phi looked visibly lighter and said, "There is much to discuss. Lyra will show you to your living quarters. If you need anything, ask the guards outside. I will instruct them to make any accommodations necessary."

Oizys sat quietly by himself.

"You are welcome to join us if you want," I said, half-joking.

Oizys grinned wide, and with crumbs still stuck to the

corner of his mouth, he said, "I live for this type of thing—a terrifying creature guarding the entrance to a crypt. I couldn't live with myself if I said no. Count me in."

His carefree demeanor shocked me, but we needed every hand we could find.

XIII

Decimal City

Opal

Sunlight beamed into the cave, highlighting the jagged rocks around us. My neck was stiff from sleeping on the rocks. Plex woke shortly after me.

"What a horrible night!" He complained.

We could hear gnashers snoring loudly above us, mixed with the thumping noise of their footsteps. Their perfumes provided a faint smell of tea and honey—a reminder of what I had lost, Tia. Having escaped their satchel, I felt I had a second chance.

"Best get moving," Plex said, holding his hand towards

me. "Here, take my hand."

"What?" I pulled away in shock. "No way!"

"Suit yourself." He responded confidently. In a quick motion, he stepped into the darkness, vanishing completely.

I shouted, "Where are you going?"

"I'm getting out of this cave." His voice sounded much closer than I thought it would be. "You coming or not?"

"How do you know where you're going?"

His head poked into the light, and his body remained a dark outline. "It feels like an instinct—an internal compass guiding me through. I like to think it's Mortimer telling me in my mind."

"You mean you can't see either?"

He shifted his body back and forth as if he was trying to decide something.

"What is seeing exactly? If I can safely make it through the cave, does that mean I can see in the cave?" He sat down on a nearby rock, seeming to consider his own question.

I turned to the wall with jagged rocks highlighted by the sun.

"You got Mortimer caught by a gnasher, after all. I think I prefer to take my chances with them."

I tested my weight on a couple of handholds and footholds. The smell of moist rock filled the air. I reached out to take hold of a wet, mossy rock. I slipped to the cave floor, landing on my feet.

Plex stepped in front of me.

"You realize the gnashers are up there?" He said admiringly.

I pushed him out of my way and attempted to climb out. Avoiding the wet rocks, I was nearing the top. My last handhold rested just out of reach. I stretched as far as I could. My arms screamed in pain. I pleaded for the rock to move closer, my fingers fully outstretched.

"Almost... Agh!" I groaned, placing my hand near my side to let it rest.

Plex sat on the floor. He leaned back casually to watch my performance. His lips were curled into a gloating smirk, as if he was waiting for my inevitable failure.

I tried to ignore him. I reached out again—stretching—fingers touching the side of the rock. In the excitement, my sturdy foot slipped. I tumbled down to the floor and landed hard on my side. I gasped for breath, but remained determined to make it out.

Every attempt I made failed—rocks scratched my arms and legs. Every impact caused a fresh bruise. Breathing heavily, sore and tired, I sat down.

"You don't give up easily." Plex said, "That's admirable, but is my intuition *really* that bad?"

"Yes... it... is that... bad." I responded between breaths.

He laughed.

"I've done this before."

As much as I didn't want to admit it, I knew. *I simply wasn't tall enough.*

"You promise you've done this before?" I asked.

Delightedly, he stood up and extended his hand.

"I do it all the time." He said confidently.

I took his hand, feeling defeated.

"Lead the way."

Despite my hesitation, Plex was surprisingly good at navigating in the dark.

"Careful of your step here," he would say. Slowing down slightly to ensure I didn't trip.

"This part up ahead is real slippery." And sure enough, a few steps in front of us was a patch of moss that nearly caused me to fall.

I was glad he didn't make any stupid remarks about how he was right. That would have been too much.

"Low ceiling." He stated, placing my hand on the lowest part to ensure I ducked far enough to get through.

"Are you sure you can't see down here?" I asked.

"I can't even see my hand in front of my face." He said confidently. "I think I've always been a part of Mortimer. He would travel to these caverns all the time. Even though I couldn't interact with him, I could catch brief glimpses from the sidelines."

Faint light seeped into the cave from up ahead. Plex's silhouette sharpened, no longer just a shadow but a figure I could follow with confidence. The jagged outlines of stalagmites and stalactites took form. A thin layer of moss softened their edges. The damp air carried hints of something fresher—a faint breeze, the smell of flowers and a promise of an open sky.

The green moss clung to the rocks in vivid patches, and the uneven ground gave way to something soft. Step by step, the weight on my chest lifted. I could see enough to walk unaided, my hand no longer grasping blindly at Plex's for guidance.

When we finally stepped out into the light, the sun's warmth washed over me, chasing away the cave's oppressive chill. I stopped. I closed my eyes and tilted my face toward the sky.

We didn't die. I thought.

"Thanks," I said reluctantly, but gratefully.

His face grew bright red as he looked away.

"It was nothing."

"No offense, but I never want to do that again," I stated, shaking off the eerie feeling of having no control.

It had only been a day since I left Countable Grove, but it felt like a week. I couldn't help but wonder what other mysterious things I would discover in the days to come. Mortimer and Plex were by far the most perplexing pair I'd met.

"Does it hurt?" I asked.

Plex shot me a look of concern.

"What?"

"When you morph into each other. Do you feel pain?"

"Oh, no. It's kind of like waking up from a dream. It's rather frustrating to move around with another individual controlling half your body, though."

"Are there others like you?"

"Near the sea, there are thousands of us. It's rare to see anyone shift and change, though. Most of them take on a specific form and stick with it. Mortimer drank of the sea long ago. Most dots there didn't believe that he had a complex part, so he left and found a cozy place in The Negative Caverns. When I showed up, everything changed."

I considered how strange it would be to wake up one day with another head attached to my shoulders. It felt like something that would be a concern, but Mortimer and Plex seemed to do just fine.

In front of us loomed a gigantic wall made of stone. The surface was smooth to the touch and impossible to climb. As far as I could see, in either direction, there was no sign to the end of the wall.

Plex was giddy with excitement. His bouncing reminded me of Wriggles scampering off to bring me seeds.

"That wall is undeniably Decimal City's," Plex said with a wide grin. "Do you know what that means?"

"What is Decimal City?"

"What?" he gasped. "You've never been to Decimal City? How long have you been around? It's only the best place on the island. Equipped with the most fun and games you could ever imagine. Entertainment for years. It's said that most people who go in never leave because it's just too good."

As if he realized his energy was revealing too much of

his desire to go, he folded his arms, and leaned against a tree, and attempted to play cool.

"We might as well take a peek inside, right?" he said, glancing at a nearby shrub as if he couldn't care less what I responded.

The shrill noise returned. I covered my ears and fell to the ground.

I must be close, I thought. *The wave is coming from Decimal City.*

Plex's body bubbled and morphed until Mortimer grew out of the side of his neck, their body nearly transparent in the sunlight.

Mortimer yelled at Plex.

"I was stuck in that gnasher for days! The least you could have done is ensure I wouldn't be the one stuck in there."

"Easy!" Plex, offended, attempted to defend himself.

Mortimer continued, "How did you get captured by a gnasher?"

"Um, well," his eyes darted around, "It's a bit of a long story."

"I think we have time." Mortimer planted their right foot on the ground and stared intently at Plex.

"Fine! I won't move either," he said, stomping their right foot.

Mortimer's face fumed with fury.

"You stupid..."

I interrupted, "Plex, which way to Decimal City?"

Like a switch flipped, joy overcame Plex.

"Did you hear that, Mortimer? We're going to Decimal City!" he said, bouncing up and down.

Mortimer remained firmly in place. While Plex bounced around excitedly, I silently pleaded with Mortimer to keep moving. His anger softened as he noticed my plea. Looking a bit defeated, he agreed to keep moving. We walked in the direction Plex pointed.

I tried to tune out their arguments and followed the line of trees near the wall. In some places, the thick underbrush was all the way against the wall. My cuts and bruises from the cave stung as leaves and branches grazed them. Eventually, we found ourselves at the corner of the wall, where everything opened to a field. A large path flowed through the center of the field, leading to the city's entrance.

"What?!" Mortimer complained. "Why don't they have this path surrounding the city?"

A line of dots had formed on the pathway that included dots of all shapes and sizes. The idea of never seeing a half-size dot before I met Tia seemed crazy to me now. What surprised me about the line was that some dots were larger than I was. The largest was six times my size.

Looking back at the wall, I could see the tallest tower inside.

That's where my friends are. I thought to myself. A nervous feeling overcame me. *What would happen when I get inside? What was I planning to do?*

The tower revealed no secrets about its contents. Black and reflective.

Was it made of glass? It looked like cubes stacked on top of one another. I shrugged off all the questions. I need to get inside first. Four smaller towers, about half the size of the central one, stood at each corner of the city. Officers stood at different levels of the corner towers, looking out over the vast landscape.

"Would you look at that!" Mortimer said, staring in wonder.

The wall above the gate was about twice the size of the surrounding walls. Nine numbers were on individual balconies near the top.

They were 3 4 0 5 2 7 | 3 6 7 from left to right. A large pillar separated the 7 and 3.

More officers were on each of the balconies under the numbers. The officers on the leftmost numbers slumped over the railing as if asleep—or bored. The closer the dots were to the pillar, the more frantic they were.

Plex was beside himself. "I can't believe we are about to go inside! All of your hopes and dreams are about to come true."

Mortimer rolled his eyes.

Stuck waiting for the line to move, we watched the numbers increase 5 2 7 | 3 6 7 became 5 2 8 | 3 6 7, then 5 2 9 | 3 7 7 followed by 5 3 0 | 4 0 0. The first three numbers remained unchanged.

"It's counting," Mortimer said aloud. "What is it

counting?"

Glad to have a distraction, I examined the numbers closely. The doors' movements were visible from our proximity. Two guards stood on either side of the gate, and one in the center opened the doors.

"Welcome to decimal city," they would say with a smile, tipping their hat. Each dot that moved inside caused the numbers to change.

"It's how many dots are inside!" I said excitedly. "Look,"

I pointed out what I was seeing to Mortimer.

"I'm so jealous!" Plex said, drooling over the dots, able to go inside.

I closed my eyes to imagine what it might be like inside. Three hundred thousand... There were more dots here than there inside Countable Grove. I tried to picture introducing myself to every dot inside. How long would that take?

Suppose I spent one second with each dot. That's three hundred thousand seconds. Sixty seconds makes a minute—that's five thousand minutes. Sixty of those make an hour—eighty-three... *It would take 83 hours to introduce myself to everyone inside.*

For the first time, I realized how few people in Countable Grove I'd met. Tia, Aleph, Al, Archy, Thorne, and five others. That's not even one percent. I spent most of my time interacting with less than one percent. I thought for a moment about how many other dots might be like Aleph. How many of them have a pet like Wriggles?

Everyone always seemed so nice in Countable Grove, but I can't say that definitively.

I thought of the rude store owner after I helped them get their bag down. *What if the majority is more like him?* I pushed the thought aside. I preferred to think that the majority are like Tia and Aleph. That's been my experience so far. *Why would this city be any different?... None of them were Tia.* My eyes were wet with tears. I turned to ensure Plex couldn't see. I didn't want to explain why I was crying, not to him.

We were getting close to the entrance now. I could overhear the conversations with the admittance officer.

"How many are in your party?"

"Just me," said a rather loud voice.

"ONE MORE!" the guard shouted.

I didn't notice until now that the dots on the balconies were the ones that changed the numbers. Those furthest away from the post were the least busy. *How long would they need to wait before changing a number?*

"How many in your party?" The guard said.

"There are four of us," a timid voice responded

"Very well. FOUR MORE"

"Next!" the guard shouted.

It was us—we were next.

Plex stepped forward.

"There are three of us, sir!" Plex said, saluting the officer.

"Three of you?" The guard eyed us up and down.

"We count bodies, not heads."

"Very well. Two of us," Plex responded, hopefully.

Plex looked incredibly nervous, considering how excited he was to get inside.

"It's OK," I whispered to him.

"You don't get it! They won't let me in."

"What, why?" Mortimer asked, irritation in his voice

He whispered. "How do you think I got inside that gnasher sack?"

"What? You decided to tell us now!" I complained. I was furious. My heart hammered in my chest. "Why do you want to get back inside?" I whispered.

"I can't explain it. You will understand once we are inside."

My eyes darted around the entrance, looking for a way out. *What did gnashers have to do with Decimal City?*

"TWO MORE," the officer shouted

I backed away, no longer wanting to go inside.

"You cannot leave once you arrive at the gate." The officer said.

As I turned to run, the shrieking noise from the tower halted me. I covered my ears and fell to the ground.

Plex disappeared, leaving only Mortimer. I watched as the crowd of dots around me split into smaller duplicates. The vines stretching out from the city swept a few of them away. The guard, unaffected by the noise, grabbed hold of me.

"No! Get Off!" I struggled against the guard.

Chaos erupted behind us—dots scattered, tiny ones darting between larger ones as guards barked orders.

The officer holding me tripped over one of the smaller dots that ran behind them.

I fell on top of them.

The brute force was enough to loosen their grip.

I ran.

A guard tackled me, slamming me to the ground.

Adrenaline surged as I struggled to break free.

"No! Get your hands off me!" I shouted, twisting.

The officer tossed me through the gate and scoffed, "Welcome to Decimal City."

I hit the ground so hard it knocked the breath from my chest. Scrambling to my feet, I turned back—only to see stone. The gate was gone, replaced by an unyielding wall.

XIV

A Leader's Remorse

Eulalia

They carved the guest quarters into the stalk of a large mushroom, similar to the hollows in Countable Grove. Scattered around the room, luminescent fungi displayed a rich assortment of colors.

"We have stocked the cabinets and provided you with all the resources," Lyra said.

She bowed and left.

There was plenty of seating, a study, and a comfortable area for sleep. I'd never seen such elegantly crafted jewels and artwork. The lights provided unique perspectives to each of the paintings on the walls.

Ellis was observing the various statues scattered around the room. She was walking a circle around one when she said, “Look at how the light shows through this one.”

Made of clear glass, the artwork shimmered, casting vibrant hues across the room; the colors each person saw depended entirely on their perspective.

Oizys found a cabinet full of clothes. His cuts and bruises peeked through his torn clothing. They looked painfully red and irritated, but healing.

“Think they will mind if I use some of their clothing?” He asked, holding up a pair of trousers.

“Why else would they have them in here?” Ronan said, jumping onto a bouncy bed. Mud and muck from his boots were draining onto the sheets.

“What’s this rune?” Ellis asked, pointing to an enormous statue of a reverse capital E.

I shrugged, “Not sure. I don’t know that I’ve ever seen it before.”

Ellis stared intently at the item, hoping it would tell her its purpose.

My eye caught a piece of parchment and a porous mushroom stalk resting upon a sturdy desk. I sat down, hoping to send a message to Countable Grove.

Despite its soft appearance, the mushroom stalk had a coarse, rigid feel, like bark. I put the tip on the paper, and as I moved it across the parchment, it left a dark mark. With some trial and error, I eventually developed an efficient way to hold the mushroom stalk.

Fellow dots. We are safe and doing well. We are pursuing The Pythagorean Crypt. It may take some time for us to overcome the obstacles. We are doing well. We hope to rejoin you soon.

~ *The Caretaker.*

My aching arm shook from the strain of writing. My penmanship was mostly legible. I folded it up and sealed it with the provided wax. I gave the letter to the guard. After a few details, they left immediately.

I dropped down on the plush bed, feeling the cool sheets against my skin. The room was quiet, with only the faint sound of rustling pages as Ellis read nearby.

"Fire!" someone shouted.

Groggy and still half asleep, I pulled myself out of bed to look out the window. Ronan, Ellis, and Oizys all looked as tired as I felt. The warm glow of flames reflected off a nearby mushroom.

Dots of The Negative Caverns ran frantically.

"Get out!" A dot shouted.

"Fire!"

The four of us scrambled to get out as flames erupted all around us.

The entire village was in a panic.

Ellis started to cough and gag, already feeling the effects of the smoke.

"The dining hall! Everyone to the dining hall!" A voice yelled.

We attempted to follow the crowd. The fire's intense brightness overwhelmed the faint light from the translucent dots, making them disappear into the rocky, fiery patterns.

Fortunately, the fire illuminated the path back to the dining hall.

I blinked, attempting to fight off the burning in my eyes. My lungs screamed in protest with every breath. Hacking and coughing, we pressed forward. Safety felt just out of reach. The heat from the flames was all-consuming.

With each breath, the steps forward became heavier.

I heard coughing and sobbing nearby. We were close to the exit.

With one final push, the four of us collapsed in to the safety of the dining hall.

Every cough felt like my lungs would burst, followed by a dry gagging. Slowly, I regained control of my body. I pulled my knees to my chest.

The bright flames fought back the darkness of The Negative Caverns. Reflecting off the wet rocks and walls, making the cave feel much smaller than I originally thought. Some areas remained untouched by light.

"Who would do this?" Dots protested between coughs.

"I need answers now!" Dux demanded.

The guards looked anxiously from one to the other.

"It must be our guests!" A voice shouted.

"Nonsense!" Phi said without a glance in our direction. "We either have an intruder, or someone here has betrayed us."

"Twenty guards, go! Search the seven bridges." Dux commanded. "The rest will remain here in case they attack. Report any signs you may see."

Like a well-trained battalion, twenty guards immediately ran to unique tunnels and entryways.

We watched the ever-consuming fire destroy the village for nearly an hour and scanned the walls for any signs of stray flames.

The search progress was easy to detect as the guard's glow splashed hues across the nearby rocks.

"There! Look there!" a guard pointed across the river.

A flickering orange flame revealed the location of a torch.

"And there!" another voice yelled.

"After them!" Phi demanded.

The torches moved in large circles, allowing the guards to catch up swiftly. We watched as the teal splashes overcame the orange flickers. With each dimming torch, cheers erupted.

As the flames died and turned into embers, we waited for the guards to return with the prisoners.

Glowing again, negative dots walked through what remained of their village.

Two guards returned about the same time. Each one's clothes and staff glowed like a magical artifact as they poked and prodded the bound dots.

"Oh!" Ellis whispered. "They use magic to interact with naturals."

The memory of Ronan flowing through Lyra yesterday flashed through my mind. There is a way for us to interact with each other.

Each prisoner held their hands up, tears streaming down their face.

"Explain yourself!" Dux demanded.

The prisoners said nothing. Their rags of clothes clung helplessly to their body. Singed and burned, ripped and torn. They sobbed.

"I'm sorry. I'm so sorry." One prisoner sobbed. "She is a monster! She made me do it. I couldn't let my friends suffer anymore. Please don't hurt me!"

With a crack, a guard shoved the blunt side of the staff into the prisoner's back, forcing them to the ground.

They clutched their broken ribs.

"Stop!" I demanded. "Do not harm them!"

Dux and Phi gave me a look.

"Allow us our customs!" Phi said coolly, then waved for the guard to continue.

"If you want my help with the crypt, you must stop. Now!" I insisted.

Phi motioned to her guard to hold.

"We need justice for our people," Dux responded. "Do

not make an enemy of us."

Ellis was at my side, cheeks dark from the soot.

"There are other forms of justice," I responded. "Your methods will only kill them or make further enemies."

Phi stared into my eyes, testing me. She pulled the prisoner's face towards her.

The prisoner grimaced in pain.

"Who is this monster you speak of?" she asked.

"Arithma," they said.

The scene was familiar. Like I'd been here before, but I knew I hadn't. I felt a flutter of confusion, then all at once, realization settled in. *This* is how Gaia was so successful. She led with destruction and passed that knowledge down to Arithma. Arithma is a soldier who follows the orders of her predecessor. As if Arithma was here, my old nervous tick prompted me to touch the scar on my hand. I stared blankly at the prisoner.

"Are you ok?" Ellis asked.

I had no words.

I am no better than her. I thought. *These are Arithma's crimes.* I glanced at the destruction carried out by her orders. It was time to be held accountable for my crimes and use the knowledge to do some good.

"Dux, Phi, may I have a word?" I asked.

Phi gave an irritated glance but stepped back.

Dux seemed intrigued.

"I've seen this before. In subjects I used to rule over. Arithma trained me."

The two prisoners looked at me.

"Eulalia?" The prisoner on the left asked, as recognition settled in.

I nodded to confirm.

"You mean to tell me you are the one responsible for this attack?" Phi asked with a chill in her voice.

"No. I am not responsible. I left that life behind years ago, but it still plagues me to this day. The fires are Arithma's crimes. Though not fully innocent, these dots deserve to have the full weight of their predicament explained."

"And you believe you can provide that?" Dux asked.

"Yes. During my training with Arithma, she took adventures on her own and wouldn't allow anyone else to take part—myself included. She would disappear for months at a time. I kept things running while she was out. Occasionally, issues arose that required me to make," I paused, looking for the right word, "*tough* decisions. One of those issues was thieves that would steal from the royal granary."

Phi impatiently looked between me and her prisoners.

"I was a fool," I said. "A thief had stolen half of our gain. Enraged, I pushed forward to find the thieves. Everything happened exactly as I instructed. We ravaged the town. We tore everything down. Dots, unwilling to comply, spent days in the dungeons."

Ellis gasped.

I could feel Ronan's angered eyes pierce my skull.

Though I had spent much time working with them, they knew little of my previous life. I noted their reactions, but pushed forward.

"My reign on the people of Origin was terrible. The look in the dot's eyes that burned this place to the ground I've seen it before. I was the source of that look in others."

"Make your point," Dux said.

"Dots in power hold responsibility for the orders they give. This dot mentioned their friends were in danger. Arithma would have committed unspeakable crimes towards them had they not complied. You said yourself, 'Unorthodox, but effective.' They are exhibiting the same logic you used to starve us and place us in an impossible situation. If you can avoid responsibility for your actions against us, then these prisoners deserve the same leniency."

Phi scoffed in anger.

Dux remained seated, deep in thought.

"A very effective argument," Quint said, stepping out of the shadows. "Having been the source of a great tragedy, I also favor mercy."

Her statement seemed to ease Phi's tension.

"Arithma holds the power to rationalize irrationals?" Dux asked in solidarity.

"She does," I responded.

"Then our efforts must remain focused on finding the secrets within the crypt. The prisoners will spend five days in darkness. They will remain under careful watch in our new guest quarters until we can remove Arithma's threat."

As the guards carried them off, one prisoner mouthed, "Thank you."

I turned to Ellis and Ronan. Oizys stood behind them both.

"I can't believe you would do such a thing!" Ronan said. His hands and arms were stiff with anger.

"You're just like Arithma!" Ellis snarled.

Their eyes, filled with fear and hatred, fixed on me.

"I can't change the past," I said. "You should know by now, I have left that life behind. I understand it may take some time to rebuild trust. I hope you will still come to the crypt. We need your help."

Quint, paying no attention to the intense silence, stated, "With Arithma's increased aggression towards us, we need to seek The Pythagorean crypt immediately."

"Will the negative dots be alright?" Oizys asked.

"The village will disperse through the caverns. It will be difficult for them, but they are up for the challenge. They will survive."

Ronan and Ellis pushed in front of me to follow behind Quint. Despite the tension in the air, I felt relief they still planned on coming.

XV

The Building of Poor Order

Opal

With Mortimer at my side, we walked cautiously down the cobblestone path. Cart owners sold merchandise to thousands of dots of varying sizes. Three immense buildings cast a shadow over the bustle of sales. The largest tower marked the location of the irregular waves that abducted my friends.

"Welcome to Decimal City," a merchant said proudly with a bow. "All your dreams await you inside."

The strong smell of berries, tea, and something I couldn't place caught my attention.

"Mmm," I said, closing my eyes. "That smells delightful."

Four merchants materialized in front of us. Each one held a unique set of items. One had a pillow with the finest jewelry I'd ever seen. Another held a plate of food—the source of the smell. The third had an impossibly enormous smile and was idly tossing some keys.

"Greetings!" the fourth dot said, "You must be new to the city. Leave magic at the door. With obli, you don't need it ever again!"

He flung two bags of coins at us. I caught mine, but Mortimer's fell to the ground, and coins spilled out. He leaned down to retrieve them.

"This is just a gift to get you started. My banking services shall provide you with everything you need." The dot said, bowing slightly.

The bag had a satisfying heftiness. I tested its weight by bouncing them slightly in my hand.

The banker was gone as quickly as he had shown up.

I overheard him give the same statement to another dot behind me.

Another merchant threw Mortimer a set of keys and pulled him closer to discuss lodging—whatever that meant.

"Imagine having twenty-two acres all to yourself." He spoke in a dreamy tone.

Another merchant with the food took a step closer.

"May I offer you our finest in dining, my dear?" He said

with a low bow. "I give you the first meal completely free."

He held a plate with varying colors of berries resting atop green leaves. I was rather hungry. I took a few bites of each item. Feeling the energy return, I was grateful for a meal.

"Thank you," I said, handing them one of my obli coins.

"Of course! Our services are always here." They bowed and stepped away.

Mortimer ran up to me, arms full of items.

"This place is amazing!" He said, "You can just ask for more obli, and they will give it to you. Look at all this stuff I've got!"

A walking stick that resembled a tree branch easily found in the forest. It had no unique appearance, but Mortimer was excited about it. He held up a new pair of clothes.

"The socks came free! They just gave them to me."

His face beamed with delight.

"Oh! I almost forgot," He said, dropping some items to pull out a large key. "I got us a place to stay."

Before I could respond, another merchant pushed their way in front of me.

"You have an eye for fine jewelry." He said, pointing to my necklace. "Might I interest you in some magical artifacts?"

A pillow highlighted a necklace, ring, and earrings. Each one polished so finely I could see a resemblance of myself in the reflection. A push sent the jeweler sprawling.

"A beautiful dress for a wonderful lady!" A tailor said, holding a beautiful green dress up to me.

A dot, holding a mirror, trailed the tailor.

The dark reflection reminded me of the building in the center—rekindling my focus.

"Let's go!" I demanded.

I pressed on, ignoring the shopkeepers. Though I attempted to appear uninterested, their advances were relentless and pervasive.

"Only five obli!"

"On sale for 10 obli!"

"Buy one, get one free."

"What is obli?" I asked.

Mortimer's face lit up.

"It's short for obligation." He said, "Sounds like you get obli first and then do the work later."

He looked around as if holding onto a secret.

"I'll simply say I cannot do the work!" He whispered and roared with laughter.

"Wait, you mean you get the thing doing nothing?"

"Basically." He chuckled. "What a crazy place! Anything you want at your fingertips."

I was skeptical.

It took a massive effort, but we eventually reached the tower's base. I looked up to analyze the structure. As if confirming my suspicion, another wave shot out from the top of the tower. I could see the wave but didn't hear any sounds. Even more peculiar was that no one in Decimal

City was affected.

No one split into smaller versions of themselves. Mortimer stayed Mortimer.

At the foot of the tower, two guards stood watch. Both had green vines covering their face and cloaks touching the ground. A palpable sense of gloom hung about them.

"Names!" One demanded.

"I'm Opal, and this is Mortimer."

They looked over a piece of parchment. Scanning it from top to bottom. With a deep, intimidating voice responded,

"Not on the list. No entry!"

"What?" Mortimer said incredulously. "How do we get on the list?"

No answer. They stood like statues.

Ignoring them, I walk by.

In a single motion, I landed hard on my butt.

"No entry into the building without a name on the list," they said.

I inquired of dots wandering nearby. All of them ignored my requests.

"How do I get inside?" I shouted in frustration.

"We're all trying to get there, sweetheart," said a provocatively dressed dancer.

Frustrated and annoyed, I collapsed at the base of a statue.

The caption read, *Argentarii — The First Mensarii.*

It was a statue of a man standing on top of a mountain

of coins, with dots climbing after him.

I stared into the reflective surface of the opal necklace. I twisted it back and forth, hoping it would provide a solution.

"Wait... That's it!" I said, standing up. "Hey Mortimer!"

He turned with a mouth full of a pastry.

"Do you know how this thing works?" I asked, pointing to the necklace.

He struggled to get the bite down.

"You're the witch. I should ask you how it works."

"Why was it glowing when we merged back in the gnasher satchel?" I urged.

He shrugged and went back to his pastry.

The magic captivated my mind. It's like it canceled us out. Like we became nothing.

Maybe I just need to recreate the falling motion. I thought.

I ran at him.

Mortimer's face changed to shock. Knowing there was no chance to stop me, he frantically put the food down.

I jumped.

The necklace glowed.

We merged.

Our bodies became nothing again. I could hear Mortimer in my mind.

"Warn me next time, would you?" He said in frustration. "My pastries are all on the floor."

"Sorry," I said. "I will next time."

Ghost-like, we moved past the guards and floated

through the doors of the building.

There were just as many merchants inside as there were outside. It was a welcome change that vendors didn't constantly try to sell me things.

Grateful for the change of pace, I stopped to admire some of the clothing hanging just outside a merchant's shop.

A sign in front read, *Equip your storefront easily with this fantastic deal. We will throw in a pair of knitting needles with every purchase of 500 units. This ensures you can produce more clothing for less, which puts more obli in your pocket!*

I imagined having a lifetime supply of clothing. *Where would they put them all?*

A well-dressed dot was wandering the center of the floor. He had a three-piece suit on with a bow tie and well-trimmed hair.

"Miss!" He said calmly. "Can I help you, miss?"

He had neatly folded his hands in front of him.

"Um... How do I go up?" I tried.

"Oh, of course." He held his arm out. "Just use the elevators to the right."

He pointed toward some metal doors with buttons on each side.

"I've never used an elevator before. Can you help me?"

"Certainly." He gracefully made his way over to the

metal doors.

"A unique form of magic powers this elevator," He said admiringly. "An ancient artifact recovered by our top scientists. We have wizards standing by that cast magic to make it work. Pressing this button will notify them you need a service. They will bring the elevator to you. Once inside, you can press a button on the floor you wish to visit. These elevators are one of a kind. Unique to Decimal City. Invented right here in this very building. Be sure to stop by the architects and engineers. It's lovely to see everything they are working on."

"Do you know what is on the top floor?" I asked.

He eyed me suspiciously.

"The top floor is for dots with the most obli. It would be unwise to go there. Is there anything else I could help you with?"

"No, thanks," I said and turned to look for Mortimer, who I found talking to a merchant. "Mortimer! We can use the elevator."

He held up his hand while intensely looking at a piece of parchment.

"But why would anyone have that many berries?" He asked in frustration. "What do they plan to do with them?"

Inscribed on the paper was a puzzle.

A basket contains the same number of rocks as it does berries. If you remove 10 berries and two rocks, there will be three times as many rocks as berries. How many berries were there originally in the basket?

"Mortimer, let's go!" I said, trying to get his attention.

He continued to stare intently at the parchment.

"This seems like a ridiculous scenario! They would just count the number of berries left and add ten. Can't you give me a practical puzzle?"

The merchant laughed. "Well, sir. If you cannot solve it, that will be 10 obli."

"Very well." Mortimer said.

He pulled some coins from his pocket and handed them to the merchant.

"Let's go!" Mortimer muttered. "What a scam—he offered me 100 items for free if I could solve the puzzle."

To reduce the number of ways he could get distracted, I rushed him to the elevator. Once inside, there were 10 buttons. I didn't recognize the numbers, and they were in what seemed like random order. Evenly spaced, the buttons formed a triangular pattern. I reached up and pressed the button furthest from the floor.

The doors closed, and the elevator lifted us into the air until it stopped. The doors slid open.

A sign in front of us displayed *The Gambling Hall*. This floor made the chaos of the merchants below feel like a lullaby. Exhausted dots stood before large machines, pulling levers that triggered loud noises.

Other dots sat around a large table, holding cards and placing obli on the table. Once a winner was determined, they would take the coins in the center. Above the dots was a banner that read, *Meet the Bernoulli Brothers today.*

A band stood on a high platform, playing music that rang through the space. The machines made their noise as well. We made our way to a window on the other side.

"I won! I won!" a dot shouted in excitement.

Coins spilled out of the machine and covered the floor. The whole floor burst into an even busier bustle of commotion.

Dots sprinted to the winner, scooping up as many coins as possible. Dots threw punches and elbows. A few guards with green vines on their faces slowly removed dots. Within a few moments, the whole place returned to a less chaotic state.

Looking out the window, I could see we were only a couple of floors up—nowhere near the top.

"Why wouldn't they put the top floor at the top of the buttons?" I asked, irritated.

Mortimer's eyes continued wandering to the machines. Blackjack, slots, poker, and craps all welcomed him to place a bet. "We could have so much obli with that!" He whimpered. "Think of all the things down there we could buy!"

He stopped by a machine with a lever attached. A massive sign over a group of them read, *Winner every 10 turns!*

"No way! Only ten turns to get a massive payout." He placed an obli coin inside and pulled the lever.

Three wheels spun wildly. The first and second wheels stopped on a berry, and the third was a jewel. The machine made a weird noise and read, "Try again!"

He placed another coin. This time, it was a rich-looking dot, another berry, and a leaf.

"Nothing again!" Mortimer groaned.

"I need to keep looking," I said, pulling away. "Please come with me."

"No can do." He said, shaking his head animatedly. "It's important that we have obli and lots of it! I will stay here and make some money while you keep looking. We'll have enough to buy this place once I'm done."

He pulled the lever again, and again, and again.

Giving up on keeping him with me, I walked towards the elevator. There was nothing about this place that made any sense. The coins they provided had no purpose anywhere else on the island. They didn't seem valuable, yet everyone wanted to have them.

A dot bumped into me, spilling a drink with a sour smell. "Whoa," he muttered, stumbling to the poker table.

I wrung out my shirt before going back to the elevator. I sensed that Mortimer's heavy use of obli was a mistake. Taking a deep breath and one last look at the gambling hall, I pressed the elevator button.

The elevator buttons were now arranged in a square

pattern instead of a triangular. I still didn't recognize any of the numbers. I pressed the second button to the right of the top left corner. The elevator sprang to life.

The Library

The room was well-lit. Plush seating comfortably filled the space. I took a calming breath. The peaceful ambiance and presence of books created an illusion of Tia's tea fragrance in the air.

Finally, I found a place free from merchants and noisy machines. I smiled as I walked toward the window.

I looked out.

An unexpected wall separated the front and back sections of the city and intersected the ominous building roughly halfway up.

The back section of the city was shadowy and still. A gnasher-sized gate stood on the furthest wall.

I was thrilled.

Maybe a way out? I thought.

I studied the dark corner for a few minutes before returning to the elevator.

I'm still not at the top.

While waiting for the elevator to return, a dot resembling the concierge from the base floor wandered into the library. He wore a collared shirt with reading glasses, which reduced the intimidation factor of the vines covering his face. He calmly gave me a nod and sipped tea as I stepped into the elevator door.

This time, a pentagonal shape of buttons rested in front of me. Giving up on making sense of it, I pressed a random button.

The elevator went down.

"Argh!" I groaned.

The Basement

I stood on a floor lined with cots, their thin mattresses old and permanently imprinted. Rustling curtain dividers partitioned the room, muffling the sounds of quiet conversations. The air carried a thick musty smell of old fabric. I could feel the tension in the room as I noticed the dots scattered across the beds, each lost in their own thoughts.

No windows.

Some read books, others played cards, and some rested with the curtains drawn. The ground shook as the wall opposite me rose, revealing three vined officers.

The room erupted.

Every dot sprinted towards the elevator.

I stepped aside.

An officer threw a bed, sending playing cards flying across the room.

Several dots were in the elevator, frantically pressing

buttons.

The guards found the dots they were looking for. They bound them with vines.

"You have not fulfilled your obligation to repay," the guard said in a lifeless tone.

As quickly as they showed up, they were gone.

They sealed the wall shut, solidifying their exit.

The room gradually went back to normal.

"Where are they taking them?" I asked, a dot standing next to me.

"No one knows. The merchants don't clarify that it will need to be paid back. Newcomers do not have enough caution when spending obli. Most of the dots here are trying to avoid the guards."

I felt an initial sense of justification. Concern quickly replaced it.

Mortimer! I thought. *I need to warn him. We need to get out of here.*

"I'm Opal," I said, hoping for more information.

"I'm Alexander." They tossed their head back as they fell backward onto a bed.

"I need to warn my friend. Do you know how to get to the floor with all the machines?"

"The gambling room? That one gets a lot. Good luck finding your way back to it. Some dots get lost in here for years before finding a way out. None of the floors match the elevator numbers. Whoever designed it made it intentionally confusing."

A chill ran down my spine.

"Do you know how to get to the top of the building?" I blurted out.

Stunned by my question, he stared in disbelief.

"You don't want to go to the top floor." He said flatly. "No one comes back once they make it to the top."

"Why don't they come back?"

"If I knew that, I wouldn't be here." He shuffled some cards while he spoke. "I once had a friend who believed this whole place was a setup to create gnashers. He believed everyone everywhere was out to get him. He got in his mind that the safest place was *actually* the highest floor because no one would think to look for them there. To this day, I do not know if he made it there or not. All I know is I haven't seen him since. Whatever's on that top floor is not good."

My breath quickened.

"Thanks!" I said, sprinting to the elevator.

"Don't go to the top floor." They shouted ominously at the ceiling.

His words permeated my mind as I waited for the elevator.

I have to go to the top floor. That's where my friends are. I stepped into the elevator with a greater determination.

The Performance Hall

The doors slid open to reveal a long, dimly lit hallway. Incense filled the air, giving a mystical aroma. Wizards in colorful robes stood on a raised stage as smoke swirled around them. Palpable excitement filled the air as the crowd reacted with cheers and gasps of amazement to the wizards' stunning feat of shrinking a colossal elephant to a tiny speck.

"Wrong floor," I said, frustrated.

A volunteer joined them on the stage. Additional cheers echoed through the space as the dot grew to the size of an elephant.

The elevator door closed again—the buttons remained in the same arrangement.

Is it because I didn't leave? I wondered, pressing another button.

Alexander's ominous warning rang in my mind—*Don't go to the top floor.*

XVI

The Pythagorean Crypt

Eulalia

The air grew icy as Ronan and Ellis clumped together, whispering to one another. My confession was painful to them.

I couldn't blame them. Al was the only person I spoke with in-depth about my past life. I believe he kept the stories to himself. The tales of Countable Grove provided a misleading impression of my goodness. I was a fallen hero to them, but it was necessary. A step I was all too familiar with. Arithma was an incredible teacher. The shift from admiration to disgust was a painful awakening.

Oizys noted the tension.

"It was brave to share that story." He encouraged. "Give it some time."

We followed Phi back to Dux, perched on a boulder near a makeshift table of granite.

Lyra unfolded the map onto the table.

"The days ahead are going to be challenging," Dux said.

"I'd say," I overheard Ronan whisper to Ellis.

I shrugged off the comment. We had more pressing matters. We needed to unite the five domains to fight the evils ahead. The leaders of The Negative Caverns were powerful allies. Supporting their needs helps us get closer to defending against the beast.

I hoped we had enough time.

"Lyra and Quint," Dux said. "Your task is more critical than anyone else's today. I need you to retrieve and protect the negative life force known as the additive inverse artifact."

Lyra bowed in confirmation.

"Ancient magic already guards the artifact," Quint challenged. "Do you think it is wise to retrieve it?"

"We need to keep it from Arithma," Dux responded. "Clearly, she can navigate the caverns. If she finds it, we lose everything."

Quint reluctantly agreed.

Dux took a slow breath before addressing the group.

"The rest of us will go to The Pythagorean Crypt."

"Might I suggest." Ellis interjected. "If this artifact is that important, why not send the larger group there?"

"A wise observation," Dux said. "The only dots familiar with the artifact's location are those going to retrieve it. They can travel swiftly and undetected. Arithma will be sure to follow if we send too large of a party."

Dux nodded in Lyra's direction. She turned on her heel and Quint followed closely behind.

Attention turned to the map resting on the table. A clear picture of the underground network of caves and caverns.

"These details are incredible!" Ronan admired.

"Aside from the location of our life force. This is everything we know about the caverns below the island," Dux instructed. "They go deeper, but we've never reached the aether source."

Ronan pointed to a place on the map. "What's The Lonely King of Seven Bridges?"

"Ah," Dux gave a somber expression. "That's a tale worthy of exploration. I do not have time today to give it justice."

He pointed to the crypt's location on the map.

"This is where we need to go," He said, sliding his finger to a nearby exit. "We will take this passage to the surface and resupply."

Phi's voice rang out in warning, "The Pythagoreans are sure to have many ways of protecting their secrets. No one knows what lies beyond The Beast of Eternal Burden. Once you make it past, proceed with caution."

A silence fell upon the group.

Ellis and Ronan continued to avoid eye contact with me.

The thought of risking our lives in silence, without a word to each other was agonizing.

Despite the growing tension between Ronan, Ellis, and I, the trip to the surface went smoothly.

Oizys used the time to help us prepare for what was ahead.

"What do we know about the Pythagorean Brotherhood?" he asked.

"Given their secrets, we know little. However, we know some things. They believed sacrifice was the most just. Medicine was the wisest. Harmony was the most beautiful. Decisions were the most powerful."

The passage forced us into a single file line.

"Bats will fly overhead. Duck to let them pass. They are harmless." Phi instructed.

An ever-increasing light brought hope to the end of darkness. The brightness sealed my eyes shut as I struggled to absorb the increased light. It didn't matter. I stretched out my arms to soak up the warmth. Fresh air flowed into my lungs, bringing a pleasant fragrance of nearby flowers. Every step provided a needed distance between myself and the mustiness of the cave.

A small stream trickled nearby. I opened my eyes.

Ronan peered into a small pool of water, using it as a mirror. In a frenzy, he attempted to flatten his wild hair.

We tried not to laugh.

"It looks good," Ellis said. A warm chuckle escaping her lips.

The air between us grew warm, and the ice melted. Though it did not last, it provided me with some needed hope.

The pleasant view of painted landscapes greeted us, with the focal point being an unnatural settlement featuring a gigantic wall and a massive tower-like structure protruding from the center.

"Look at all those Gnashers. Did they build that?" Ellis asked, pointing to the moving boulders in front of the wall.

"What are those towers inside the walls?" Oizys asked.

Dux responded, "I worked closely with dots in the past that presented blueprints for a tower that had a similar appearance. They called them buildings."

I felt an unease that was hard to place.

We rested in the sun for a couple of hours as we resupplied. Once satisfied with our rations, Phi and Dux led the way. We walked across multiple fields to get to the next cave.

The cave's entrance was about half our size and required us to crawl on our stomachs.

"I will go first," Dux said. "I will call once I make it to the other side."

I shivered at the thought of the unyielding darkness.

"How will we see?" I asked.

"This cave lets in a reasonable amount of natural light. Seeing should feel relatively normal."

Not wanting to leave the sun, I was one of the last to go through.

Ellis and I sat in silence as Ronan made the journey.

Ready for the silence to end, I crawled through to the next area. A breath of relief escaped me as I noticed a light streaming in from above, illuminating the cavern.

Eventually, the way opened up to a mid-sized cavern. It wasn't as large as the mushroom-filled village, but it was a good size, and we didn't need to be concerned about our heads hitting the ceiling anymore. A set of stairs flowed down to a dirt floor where the crypt rested in the center. Four majestically tall pillars rose from each of the four corners of the crypt, and one more marked the center. The top of each pillar included flames to offer light to the surrounding area. The light reflected off the shifting ground surrounding the crypt where the beast slithered through the dark moat.

"The Pythagorean Crypt," Phi said ominously.

A thin path provided a way to the moat of beans.

I felt scared. After taking a deep breath to calm down—I glanced around the group.

Ronan stared, in quiet contemplation, at the task ahead of us.

Oizys, surprisingly calm and unconcerned, leaned against a towering rock.

Ellis secured her artifacts and medallions around her waist, gearing up for a fight.

Phi stood next to me—staring at the moat. Fear was evident in her eyes.

Dux walked towards the nearest rock and sat down.

"Phi and I will wait here for you. Unfortunately, we will not be of use."

I stood tall, trying to project an air of confidence I didn't feel.

"You three ready?" I asked.

"Just a minute," Ronan said sternly. "Can you see a pattern?"

"I agree," Ellis stated. "This is the only time we will have a clear view of the whole crypt. We should spend some time here before proceeding."

Ronan pointed just ahead, "There only seems to be one beast. What if we approached while it was on the backside?"

"I've been here a few times now," Phi answered. "The beast is much quicker than it seems. You will not have enough time."

"A few of us could distract it while the others move into the crypt," Ellis suggested, eager to get moving.

"It's too risky." I responded, "I don't want to lose any of us. Our best chance is to have as many of us in the crypt as possible."

"Do you hear that?" Ellis asked.

She took a few steps closer, descending to the crypt. The group remained silent.

"I hear nothing," I responded.

"Quiet!" she whispered.

An eerie silence fell over the group as we listened intently.

"There is a rhythm to the beast. As it moves around the moat, it sings." She said.

I still heard nothing. The place seemed quiet—eerily quiet.

Ronan jokingly jabbed, "Are you sure you're not just hearing things?"

Paying no attention to the jab, she said, "I need to get closer." And started down the staircase.

"Don't step on the black areas. They are traps set to scare off intruders." Phi shouted down to her.

Feeling nervous, I sprinted to catch up to her. Sunlight from the rocks above illuminated the thin path. We stopped well before we got near the damned souls trapped inside the beans. A new sound replaced the tapping of footsteps.

Faint whispers.

"Come closer," they begged.

"You feel lost. Come, let us provide you with a way." The voices were warm and inviting.

The beans are dangerous. I reminded myself. *Don't get any closer.*

"I think I can hear it, too," I said.

"The vibrational hum?" She responded.

"What? No. The souls whispering," I said.

"I hear those too. This is different. It's like a vibration of aether. A feeling of safety and security."

"I don't think it's a good idea to trust that feeling," Ronan said, catching up. "Nothing good could come from this place."

"Those voices are terrifying," Oizys said, joining the group.

"Yeah, they are." Ellis agreed. "This is different, though. It just... feels right. I'm not making this up. Literal vibrations are coming from inside the temple. It feels like a gemstone in my hand, but a deep vibration from my feet to my head. It's coaching me on where to step and where to go."

A few figures of hardened clay, women and men in flowing robes, stood just ahead—some frozen mid-stride, others caught in a timeless moment of terror.

"Release me," a voice begged.

I tried to block them out.

"Please free us!" more voices pleaded.

"You can help us!" they insisted.

I can help you? I questioned in my mind.

"Yes. Help us!" they whispered as if hearing my thoughts.

I thought of the prisoners brought before Dux. *Were the damned souls in the beans unjustly trapped? Can I save them?* I felt a pull towards the beans. I needed to care for them.

Oizys interrupted my thoughts.

"I think I feel it, too," he said. "I can feel the vibration.

Now that I know about it, it's hard to miss. I'm not sure how I didn't recognize it before."

As the snake guarding the crypt slithered past, the ground trembled beneath me. I observed the beans reaching up, towering over me before descending back down. The snake's massive body matched the crypt's size.

Fear gripped me as I traced the scar on my trembling hand, hesitant to move forward. The sounds of the rustling beans intensified as the beast navigated through them.

"Free us!" they demanded. "Pain. Horrible pain. Please free us!"

Their voices echoed in my mind—one after another. It was hard to focus on anything else. Distracted and trying to keep up with the group, I ran into a rock and fell to the ground. The three of them came running back.

"Are you alright?" Ronan asked.

"Yes. Sorry, I am a little distracted by those voices." I said.

"Try to pay attention to the vibrations," Ellis instructed. "The voices get quieter when you feel the vibrations."

The whispers drowned out her voice.

"We will always be with you," the voices promised.

"There will never be cold air between us."

I had done so many things wrong. There were many dots I hurt. Ellis and Ronan knew this now. There was no forgiveness for my mistakes.

If I deserve to be free, so do they. *I can free them.*

"Yes. Come!" the voices encouraged.

With each step to follow the group, the voices sang louder and louder.

"Prove yourself to us. You are better than this! You are good. Help us get free."

In a trance, I walked towards the moat.

You're right! I agreed. *I need to prove I am good and show Ronan and Ellis I care. Prove that I am not Arithma.*

"Prove it!" They demanded. "Prove you are better than Arithma!"

I ran to the edge of the moat.

"How can I help!" I yelled.

My eyes searched the dark green vibrating beans. They morphed and changed, shifting shapes and figures until they stood tall before me. They formed what looked like a gigantic snake—staring at me. Fangs the size of my body jutting out of its mouth, ready to strike.

"You have come to save the damned souls," it hissed.

I stood confidently and declared.

"I have."

As the words left my mouth, fear returned.

What am I doing? I thought. *How did I get here?*

As if the snake could read my mind, it hissed, "You're too far in. You cannot get out. Here are those damned souls. Save them, if you can."

The snake had slithered around me, creating a circle of beans.

The whispers were now shouts of laughter.

"Murder!" they screamed, encouraging the snake.

"Kill her! Do it now!" They sang together.

I had nowhere to go. I was stuck.

"Don't touch the beans!" Ellis's voice rang through.

The snake reared back to strike.

I jumped out of the way just in time, nearly sliding into the pit. Before I could get my bearings, the snake was ready to strike again. My mind raced, trying to find a way out. I winced, imagining what it would feel like to have the sharp fangs dig deep into my skin.

"No!" Ronan shouted and dove into the path of the snake's bite.

As its fangs engulfed Ronan's body, Oizys pulled me to safety.

I stared in disbelief.

Tears filled my eyes.

I killed him. I thought. *Ronan is dead because of me.*

Ellis shrieked and fell to the ground, sobbing. The snake stopped its attack. It studied us patiently. The three of us cried over our fallen friend.

Ronan was no more. That kind of pain allows irrational thoughts to surface.

Ellis stood up and ran at the snake.

"Take me too!" she shouted. "How could you! He was good. You are only supposed to take the souls of the damned."

She hurled rocks at the beast. Each flowed through the snake's body and connected with the ground on the other side.

The snake observed, studied, and watched.

Ellis collapsed to the ground, screaming. The serpent's hiss interrupted her sobs and pleas.

A great sacrifice has been made.

To you, with rational thought and peace of mind, my secrets are to behold.

There is geometry in the humming of the strings.

Trust the rhythm within

Remember, dear child,

There is music in the spacing of the spheres.

For there are further dangers within.

The snake withdrew, leaving a path for us to move forward.

Tears blurred my vision. I stared at Ellis.

"Why!" she screamed.

"He was good," she pleaded.

Between convulsing sobs, I crawled towards her.

Please don't push me away. I thought.

Without hesitation, Ellis pulled me into a hug.

The two of us cried together. It wouldn't bring our lost friend back, but it was necessary. This was a strength Arithma would never understand. She would declare it as a weakness and turn away. However, I know the truth.

Arithma cries, too.

Emotions are a part of life. Avoiding them in front of others is more cowardly than running from a battle. It's much harder to be vulnerable with others than it is to hide in fear of further pain.

Today, my friend Ellis was brave. Today, she won a battle that fierce warriors frequently lose. She was vulnerable. My brave friend Ellis.

XVII

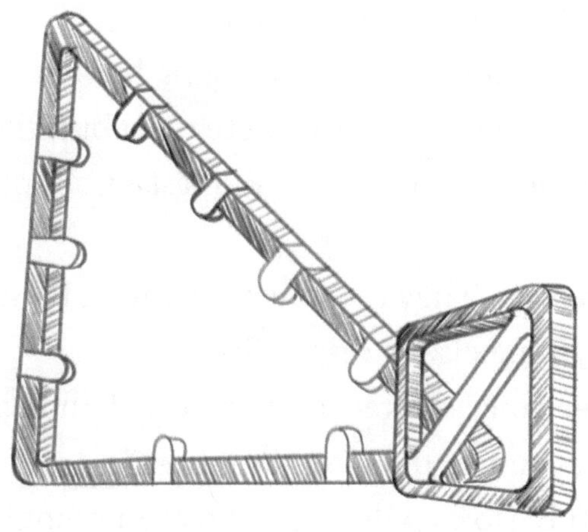

The Secrets Within

Eulalia

Something held me in place—not bindings or ropes. A dread. Deep in the pit of my stomach. My eyes were sore from crying.

A great sacrifice has been made. The snake's voice rang in my mind. Phi mentioned the brotherhood valued sacrifice. I never considered the sacrifice was a requirement to enter.

Why was it, Ronan? Did Phi know this would happen?

Oizys approached us with a burdened look. "We can turn back," He suggested. "Nothing inside is worth losing more of us."

Eyes red and puffy, Ellis drew a deep shuddering

breath. "I think Ronan would have wanted us to keep going."

I wiped my eyes on my sleeve. The dread was still heavy—Ronan's memory propelled me forward.

Oizys and I helped Ellis to her feet.

The first few steps were challenging but became easier with each one. The bean voices were quiet but still present.

"Why did you go to the pit?" Ellis asked, a hint of anger in her voice.

"It's the voices. They took over my mind. I just couldn't focus on anything else." I responded.

"Oh." Her voice shifted to concern. "I thought it was strange. Can you feel the vibrations now?"

I felt numb. How could anyone feel anything?

"I must have some loose screws after all these years," I said jokingly to lighten the mod.

No one laughed.

Ellis's gaze fixed on me. Something familiar glimmered in the corner of her eye. She didn't want to lose me, too.

"The guardian said we need to trust in the vibrations." She said, "I'm not sure it's a good idea for you to continue with us if you can't feel them."

"That's reasonable," I responded. "I will continue with you as far as I can. Let me know when the vibrations say I should go no further."

I pulled out a few berries and cast a quick duplication spell.

"Keep your energy high," I said, offering them to the

others.

We walked up the steps and into the crypt. The inside was stunningly beautiful. Intricate geometric tiles and patterns adorned the floor. The engravings illustrated a musical. Wave patterns flowed around the room, up the walls, and out the open ceiling.

Standing in the crypt observing the patterns filled my body with a vibrational energy. This must be what Ellis was feeling. I recognized the aether pulsing—we stood on the largest magical artifact I'd ever seen.

"It's a magical gemstone," I said. "A temple, of sorts, to increase aether sense and make magical spells more powerful."

I looked at Ellis with a new understanding.

"How could you feel it without touching it?" I asked.

Ellis sighed as her shoulders dropped to release the burden she was carrying. "You can feel it!" she said. "The crypt's atmosphere resembles the pulse of a gemstone. What I felt outside was different—a vibration, but a bit more like a wave from toe to head. Like a deep vibration."

Oizys walked to the lectern in the center of the room. Around him was a circle engraving on the floor featuring fifteen triangles. Scattered throughout the room were similar triangular shapes. At the heart of it all, there was a triangle that bore a striking resemblance to the grand chandelier hanging above, suspended by a few ropes.

Engraved on the front face of the lectern was the equation: $a(a) + b(b) = c(c)$

"Look at this," he stated, holding a clay tablet.

Unmistakably, the tablet was the Plimpton 322 artifact. The Babylonian tribe used it to enhance their aether connection. *What was it doing here?*

Oizys shrieked and attempted to drop the tablet, but it had fused to his arm. The clay artifact was slowly encasing his arm in clay.

"We are not the first to get here," Ellis noted. "Those statues just outside the temple were a warning."

"Oizys, does the tablet have a collection of three number sequences?" I asked, recalling in my studies that they built tablets like this to help them with construction.

"Yes. 3 4 5. 5 12 13. 7 24 25. There are fifteen sequences in total."

"Those numbers are called Pythagorean Triples. They are special solutions to the Pythagorean theorem on the front of the lectern. 3 times 3 is 9. 4 times 4 is 16. And 9+16 is 25, which is also 5 times 5."

"Why are they on a Babylonian tablet?" Ellis asked. "They lived long before the Pythagoreans. Wouldn't that make them the Babylonian triples?"

"Maybe we should save the philosophical questions for another time," Oizys said. "Can we focus on getting me out of here?"

I noted the many triangles around the room. I walked over to the smallest one and picked it up to examine it closer. It looked like the triangle in my hand would fit inside the smallest triangle on the floor.

I dropped it in—perfect fit.

Sand started filling the space.

"Look on the floor. There are 15 triangle engravings. Find the others around the room and place them in the corresponding spot."

We scrambled and worked together, slotting pieces into the engraved mosaic. Initially, we could move around with ease, but the deeper the sand, the harder it was to walk.

I slotted one of the larger triangles 16, 63, and 65 notches.

"Looks like we have three left!" I shouted.

"I have two of them," Oizys said, stumbling to his knees and pulling a heavy set of two larger triangles.

The clay had completely encompassed the left side of his body.

I sprinted to help.

"Stay out of the sand. I've got it from here." I said, pulling them from his grasp.

Ellis and I placed the large triangles.

The sand was now at the edge of the circle.

I had to move my feet quickly to stay on top of it.

Unable to move his feet quickly enough, Oizys collapsed with a startled cry.

"One more to go!" Ellis shouted. "Where is it?"

"The chandelier," I said, dancing in place to stay above the sand. Instinctively, I reached for my knife and tossed it to Ellis.

"Cut the rope! I'll help Oizys," I said.

Only his head remained uncovered.

After finding her footing, Ellis swiftly jumped and cut the rope in one move. The chandelier fell to the ground with a loud thud.

We hopelessly waited in silence.

A deep vibration shook the temple, not like an earthquake, but more like a beat. The sand stopped flowing, and the center receded into the floor. The flow of sand pushed us onto the platform. Oizys remained covered in clay.

He could not move, but was still alive.

"I think the clay stopped," He said in a strained voice.

Relieved the immediate danger was over, the three of us rested as the platform lowered into the chamber below.

"This clay is heavy," Oizys said in a labored breath. He chuckled between gasps.

"What's so funny?" Ellis asked, missing Ronan's humor.

Between breaths, he responded, "Surprisingly, this isn't the worst situation I've been in."

"What?" I said in protest. "Worse than this?"

"Yeah, it's all part of the adventure. A bear chased me into a cave after I got stuck on a mountain. I climbed into a high cavern where it couldn't follow me and waited for him to leave. Days passed before the bear finally departed."

"What's funny about that story?" Ellis asked, giggling out of discomfort.

"In the moment, the experience is terrifying, but time is an incredible comedian. I was stupid to climb that mountain by myself, but I did. Yet, I lived. I've got a good feeling this isn't the end for us. We'll make it out of here. We might even laugh about it someday."

The thought brought a brief smile to my face. The optimism provided a light in a dark situation.

"Without Ronan," Ellis said dismissively. "I don't know if I will ever laugh about this experience."

We were now at the heart of the crypt.

"Nowhere to go but forward," Ellis stated, examining the dark space we entered.

"If you need anything, you know where to find me," Oizys said sarcastically.

The light beamed from above, spotlighting a large stone slab before us. Inscribed at the top of the platform were the words.

Herein lies the secrets of the Pythagorean Brotherhood

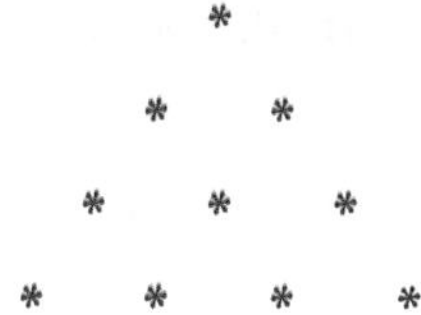

Flames erupted on the walls, lighting the room. Nine other platforms rested in a triangular pattern. Between the platforms was a dark moving floor, and the whispers of damned souls.

"Please free us," they pleaded.

I was grateful I could feel the vibration this time. It

hummed through my body. It provided stability and reassurance.

The stone slab behind us contained various strings pulled tightly. I reached out and plucked the largest one.

A deep tone rang out.

"Look! The platform next to us is vibrating!" Ellis said.

It was directly to our left. Marking one of the three corners of the equilateral triangle.

The snake's voice hissed in my ears as I remembered their words: *there is geometry in the humming of the strings. There is music in the spacing of the spheres.*

The string next to the one I just plucked was about half the size.

With a tug, Ellis plucked the string.

A higher pitch vibrated a different platform. Each of the five strings we plucked caused a unique platform to vibrate.

"They all have a similar sound, like the same note," I observed.

"Each string to the right sounds higher than the previous one," Oizys noted.

Ellis stared at the musical strings—lost in deep thought.

"What are those metal things?" Ellis asked.

Behind the strings was a set of metal edges that protruded from the wall. Each string had three metal pieces in total. A piece marking the top and bottom pulled the chord taut. Each string had a metal piece in the center, just short enough that it didn't quite touch the string itself.

There were notches engraved at different ratios of the strings. The metal pieces would fit inside the notches and create a new tension on the chord.

Ellis moved the metal piece of the largest string to the center—cutting the string in half. Together, we plucked the top and bottom portions.

The same note. The vibrating platform was the same as the string, just to the right.

We used the same method on the other strings and observed a similar effect. The fifth and final string didn't have a chord next to it. Cutting it in half caused a new platform to vibrate.

Ellis leaped with excitement.

"It's like we just created an imaginary string!" She declared. "We need to get all the platforms vibrating at once."

"Sure, let's try it." I encouraged. "There are seven other notches. What if we use the notch halfway between the center and the top?"

"The longest string is the easiest to move. Let's try it on that one," Ellis suggested.

I grabbed the metal piece and slid it against the string towards the desired location. A shrill sound echoed around the chamber. I let go of the metal piece and covered my ears.

The damned souls' voices rang louder.

"Please make it stop. It hurts us!" They begged.

The platforms had lowered slightly towards the beans

below. Once the sound stopped, the platform returned to a safe distance.

I let out a deep breath while awkwardly looking at Ellis.

"Don't move it while it's connected to the string," I said, knowing it was apparent but unsure of what else to do. I shivered to let out more nervous energy.

"Let's try that again." I placed my hand on the metal piece. I pushed it back to remove the tension. Then, I slowly moved it into position—one-fourth from the top. I plucked the longer portion of the string—three-fourths of the original.

A new platform moved.

We tried other ratios. One-fifth, two-fifths, three-fifths. Each one had a slightly different effect but resonated well.

From largest string to smallest, we adjusted the metal pieces, shifting them until they sounded good together. One-third, one-fourth, one-fifth, three-fifths, one-seventh. This gave us 10 unique sounds matching 10 unique platforms.

"Ready?" Ellis asked.

I never considered what would happen if we got it wrong until now. *Would the platforms give way? Would the ceiling seal shut?* I let out a deep breath. *There's only one way to know for sure.*

"Ready," I responded.

We strummed the top and bottom of each chord.

Every platform vibrated.

The beans flowed out of the room.

Nine of the ten platforms lowered the rest of the way to the ground. The tenth marked the tip of the triangle and remained fully extended. This new perspective revealed an inscription on the side of the tenth platform.

Welcome to the brotherhood! We entrust our secrets to you, dear Pythagorean!

Set back into the wall was a metal artifact.

I approached slowly and pulled it from the shelf.

A surge of aether energy filled my body. It was perfectly square and fit inside the palm of my hand. Like my necklace, it provided a calm radiating energy.

The center had a perfect line connecting two opposing corners and was pure gold. There was something significant about the diagonal, though it wasn't obvious.

"We did it!" Ellis cheered.

"Nice work!" Oizys declared. "Any obvious way to get me out?"

"Working on it," Ellis said, taking the square from my hand. "What is this?"

I felt the aether energy leave with the artifact.

"Another puzzle?" I suggested. "I'm not sure what it means, though. Let's look around in case there is more that we are missing."

She handed the artifact back to me.

I secured it in one of my coat pockets.

The walls were bare. A few right triangles rested around

the room.

Ellis examined one. It was a little bigger than her palm. She secured it to her necklace.

A set of stairs were on the other side of the musical stone slab.

"I'm not sure we will be able to carry him back without Ronan," Ellis said, her eyes filled with tears.

"Let's see what we can do," I suggested.

Ellis stood at his feet, and I was at his head.

We bent down to pick him up.

The triangle artifact dangled from Ellis's necklace towards Oizys. Once we touched him, the artifact glowed brightly, and the clay closest to the artifact melted away.

Ellis held the artifact closer to the clay areas that weren't melting away. Soon, Oizys lay motionless before us, completely freed.

"You did it," he said, sitting up in excitement.

I considered all that we knew about the artifact. Everything we learned about this place. The Pythagorean Brotherhood entrusted us with their secrets. Something obvious lurked within these walls—something we were missing.

"Of course! That's it!" I said. "Their magic was just an illusion. Not real. It's not based on actual aether energy. The brotherhood was a bunch of liars and thieves. The clay tablet proves the theft of the Pythagorean triples!"

Ellis could not control her excitement. She paced the room as she followed what I was saying.

"You're right!" she said. "Thieves. That means the beans are another type of trick. The beans are an illusion!"

"Yes!" I agreed. "There's got to be something here that removes the illusion."

We searched every crack but found nothing.

"If we got here once, we can do it again," Oizys suggested. "Let's get Phi and Dux. Maybe they will know what to do."

The three of us walked out of the room.

I held the square artifact in my hand—staring, hoping it would reveal its power. Within moments, we stood at the edge of the bean moat.

The voices were gone.

The serpent rose.

It hissed, "I see you have found their secret. A magical artifact that undoes their brotherhood. What do you intend to do with it?"

The square glowed brightly, casting a shield of light around us. I held it up higher to examine it. It surged and vibrated in my hand. It felt good. I felt a renewed sense of energy. Like the artifact was charging my body.

"We will use it to free a friend," I said confidently, unsure of how we would use it.

"I am the guardian of this crypt. That artifact is not to leave."

"We can bring it back when we are done." Ellis begged.

"Transgressors!" the beast hissed.

Dark souls climbed out of the pit. Each carried a heavy

chain, dragging it along in the dirt. Distortion marred all their faces, hands, and legs. As they came closer, the light from the artifact grew brighter.

None of the dots could cross the sphere barrier of light. We huddled together in the energy bubble created by the artifact—protected.

I held it higher.

The snake reared back to strike.

We felt the force of its attack as its head slammed into the edge of the force field. I lost my balance and dropped the artifact. The light went out instantly.

We were unprotected.

I fumbled to grab the artifact but caught it just in time.

The snake struck again.

I clutched the artifact tightly.

Dark figures caught in the renewed field lay on the ground, bodies contorting in strangely. Screams of pain echoed across the cavern walls.

The dark vines melted away, leaving behind motionless dots resting on the ground.

The snake kept coming at us. Strike after strike. There were no signs of slowing down.

As if the thoughts connected simultaneously, Ellis and I made eye contact.

"We can save him!" we said.

Between snake strikes, we placed down the artifact to let more of the cursed dots move closer. We picked it up to create the force field. They shrieked and shriveled in pain.

With each wave, the snake got smaller.

Ellis and I took turns picking up the artifact as it was getting heavy to carry. Both of us were exhausted.

"No!" shrieked the snake. "How can this be? You are rational, and you still seek the destruction of the Pythagoreans."

It slithered around helplessly as we continued.

It was hard to breathe. I crumpled to the ground.

"We need to move across the moat!" Oizys shouted. "You cannot keep this up forever. We should move quickly before the artifact loses power."

"We cannot just leave them all here!" Ellis said between breaths. "I won't stop. Not now. We need to find him!"

Ellis alone, stood strong. Sweat beading down her face, she fought the mob back. With each passing wave, the snake shrank.

I looked around for anything we could use to amplify the artifact's powers. There were hundreds of dots surrounding us. That's when I saw him.

Ronan—cursed and walking in chains.

"The chains!" I yelled. "Follow the chains to the source!"

Ellis heard me. She took a step and collapsed on the ground.

The artifact flung from her hand, bounced along the ground, and came to rest just before the beast.

Oizys sprang to action, sprinting towards the artifact.

The cursed dots were closing all around us. He ducked and dodged with incredible speed and agility. Narrowly missing each grab.

The snake reared back to strike—Oizys dove.

A sphere of light sprang out, encompassing the snake's head. The head disintegrated into dirt. Its body flopped to the ground and lay motionless.

Shaking, Oizys stood up. Holding the artifact high, he made his way to the source of the chains, into the beans. Light crawled through the cracks and creases. The beans now completely covered him.

I hoped he would remain unharmed.

A surge of explosive energy sent beans flying all around.

The chains holding the dots broke free.

I watched Ronan fall to the ground, shrieking in pain.

The wails and screams of cursed dots echoed against the cave walls.

Oizys emerged from the beans unharmed, holding the artifact.

We made it. We were safe. I thought.

Motionless bodies disintegrated to dust. Others remained.

I ate a few berries to restore my energy and crawled to Ronan's side.

"Wake up!" I demanded.

I placed a few drops of berry juice in his mouth.

Oizys and Ellis joined me.

Ellis slapped his face.

"Get up!" She yelled.

Ronan sat up, gasping for breath.

"What happened?" he asked.

A new atmosphere took hold. The light beaming in from above felt warmer, as did the slight breeze. We rested on a pile of motionless bodies, yet somehow, peace filled my heart. The terrifying events of the crypt were over. Ronan was alive. The remaining bodies were proof we could fight back against Arithma.

Dux and Phi were soon at our side.

"You did it!" Phi said, stepping carefully to avoid the beans.

"What was their secret?" Dux asked hopefully.

Before I had a chance to answer, Phi shrieked with delight.

"Agatha!" she shouted, sprinting towards one of the still bodies. Tears burst down her cheeks. She sobbed and carefully rested Agatha's head—shifting in unpredictable patterns—in her lap.

"I've waited so long," Phi said.

The aether energy took hold instantly.

Agatha gasped and stood abruptly, ready to fight.

"Agatha," she said timidly, gently touching her arm. "It's alright. My friends saved you."

She turned abruptly, pupils wide with fear.

"Quint? Where is she?" She glanced around the space frantically, looking for Quint.

"Quint is alive," Phi responded. "But not here."

"Lilly? Marve? Where are they?" She stared into Phi's

eyes.

"I don't know. Maybe they are here, too."

We spent the next few hours helping the resting bodies recover from the curse. Most of the remaining dots woke quickly—just as frightened as Agatha. Each one required a brief explanation of the events. A few dots dissolved into dust on contact and blew away in the slight breeze.

As the still bodies shrunk to fewer numbers, the realization settled in. Lilly and Marve didn't make it. Phi and Agatha sobbed. We allowed them a few moments of grieving.

I pulled the square artifact from under my cloak and offered it to Dux.

"This is the secret," I said. "It doesn't look like much, but it's responsible for undoing the magic of the brotherhood."

Nothing happened.

"The Additioneer's fortress is close to here," Dux said, analyzing it from a distance. "They have experience with curses. I will touch it if they deem it safe."

A shrill noise filled the cave. The group frantically looked for the source.

"What have you done?" Arithma screamed.

Thrall and 5 others followed close in step.

Arithma grabbed and threw one of the recently freed dots into the pit. She tossed another in and another.

Her anger-filled eyes met mine.

"You pitiful child!" she shouted

"It had to be done," I yelled back.

"After them!"

Thrall and the others, under the curse, sprang to action. Their movements were swift and difficult to predict.

The two dots Arithma threw into the pit climbed out unharmed.

"At least she cannot use the beans to curse anymore," I said.

Arithma clawed at the harmless beans. Screaming in horror.

"This is the one responsible for the captivity of those dots?" Agatha asked.

"She is," I said.

The unpredictable patterns of her head shifted faster as determination filled her eyes. She stood firm and in a surge of power, vines sprang from the ground, stopping each of the henchmen in their tracks.

"This way," Phi urged.

The crowd of rescued dots followed her.

Dux threw a handful of pebbles. Each one grew to the size of boulders as they rolled forward, ready to crush anything in their path.

Thrall broke free and pounded one of the rolling rocks into dust and continued his pursuit.

"Time to go!" Dux said, turning to run.

"What about Agatha?" I asked, running alongside him.

"She can handle herself," Dux responded.

Unwilling to argue, I continued to run until we caught up to the group.

"It's going to get dark," Phi yelled. "Hold hands. I will ensure you make it to the other side."

A dust storm exploded around us. We grasped for each others hands until everyone was holding on.

Single file, we sprinted into the darkness of another cavern.

"Dim your light," Dux urged each of the transparent dots in the group.

We duplicated clothes until we had enough to fully cover them, plunging us into darkness.

Too fearful to speak, we tiptoed through the dark space. Arithma's angered screams echoed though the cavern.

"Will Agatha defeat Arithma?" Oizys asked.

"Not likely, but she will hold them off long enough for our escape. She is strong and resourceful. I do not think they will capture her."

The musty smell of moist dirt was all I could focus on. The hand I held was clammy and cold. Occasionally, a whisper would pierce through the tip taps of footsteps. We pressed on.

XVIII

City Breach

Opal

With his three-piece suit and bow tie, the concierge on the first floor acknowledged me with a bow as the metal doors opened.

Stay in the elevator. I reminded myself, giving a nod in return. I studied the buttons, my fingers tracing their smooth surfaces.

"This may take some time," I groaned.

I visited about half the floors before finding the gambling hall.

"Yes," I shouted in excitement, jumping out of the elevator.

The smell of spilled berry juice lingered in the air from my last visit. The typical chaos of the floor made it hard to focus. Machine lights flashed and made random noises. I rounded the corner to find Mortimer's machine—*empty*.

"On a roll!" Mortimer's voice sang. The source was coming from a cluster of dots huddled over a game. I sprinted towards them.

"Hey!" a few of them complained as I plowed through.

The crowd erupted with shouts as Mortimer rolled the numbers he needed.

I pushed my way through until I stood next to him.

"Opal!" His eyes glowed brighter at the sight of me. "I just won 700 obli!"

"The obli needs to be paid back," I shouted.

A sudden gasp reverberated loudly among the group, followed by a deafening silence.

"That's preposterous," a dot near me whispered.

Mortimer's face gave a contorted, strange look. One eyebrow raised in suspicion.

We exchanged glances for a lengthy, uncomfortable moment until laughter erupted among the group.

"There's no way!" Mortimer giggled. "Watch this."

Turning to the dealer in a stylish suit, he politely requested another five hundred.

The dealer bowed slightly, placed another five hundred on the table and slid it over to him.

Mortimer shot me a glance.

"See. They give it out." He proudly gestured to the table.

There was a loud crash as the wall on the far side opened and pushed a machine over. Dots jumped out of the way and cleared a path between Mortimer and the vined-faced officers.

An authoritative voice echoed over the noise of the space.

"Mortimer!" an officer said.

My heart sank to my stomach. *I can't lose another friend.*

"Your attendance is required. Come with us." They commanded.

"Run!" I shouted.

He was trapped in indecision and glanced from me to the officers. The crowd of dots dissipated, providing no resistance.

"Let's get out of here." I tried again.

He let out a high-pitched shriek and jumped over the table.

The officers continued at a consistent pace, swiftly gaining on him.

To avoid capture, Mortimer climbed on a nearby machine. He jumped from one to another as the guard's bindings flowed through him.

Before I could plan, he slipped off a machine and fell to the ground.

The guards caught up to him and bound him in vines.

"What is this about?" He asked. "I'm innocent!"

He scraped and clawed at the floor below him, grabbing hold of anything he could. He threw a few coins at the officer.

Unaffected by the feeble attack, they bound his hands and gagged his mouth.

Enraged, I tried to move, but fear kept my feet glued to the floor. *What can I do? I was nothing compared to the officers.*

A deep vibration filled my body.

The entire building shook.

"It's another attack!" the guard shouted.

They dropped Mortimer, who remained ensnared by their vines.

The window furthest from the elevator shattered—spraying glass pieces everywhere. Fur and feathers filled the edges and corners of the gap as animals flooded the gambling hall.

I stared, dazed and confused—knees locked in place.

A mouse the size of Wriggles picked up a nearby dot and scampered back out the window.

"Don't let them take anyone!" shouted an officer. Vines sprang for their arms and legs as they tried to close the gaping hole.

Another officer pushed the animals back with a wall they created from floor to ceiling. Their efforts were useless.

Chipmunks and squirrels chewed through the vines. They would run around the wall. Various birds were flying in and out, snatching dots along their way.

A few officers bound a couple of critters and pushed others back out the window.

My legs finally cooperated, and I sprinted towards Mortimer. I ducked and dodged fur, feathers, and animals.

The guard secured Mortimer again and dragged him off.

"No!" I yelled.

Mortimer looked at me with pleading eyes.

Before I could get any closer, an owl swooped in and picked me up. The owl trapped me in its talons like a cage.

Helplessly, I stared in horror as I lost sight of Mortimer.

"Let me go!" I screamed.

My stomach lurched as the owl swooped down low to the ground.

The bird dodged between merchant carts and officers. They picked up three dots in their other claw and flapped their wings.

I fell to my stomach, unable to hold myself up as we gained altitude. The chill of the air flowed around the back of my neck. My body shivered in response to the cold air.

The owl leveled out, giving me a clear view of the building. Hundreds of animals were retreating—dots in their mouths or talons. The tall, mysterious building of poor order had shattered windows on every floor except the top. An eagle vigorously clawed and pecked at the unrelenting windows. There were no signs of its attempts to break in.

Every kind of animal filled Decimal City.

A bear was fighting off the guards near a destroyed section of the wall. It slashed mercilessly, cutting down officers caught in its claws. The flood of animals flowed a safe distance around the bear and out to the forest beyond.

My teeth chattered loudly. My friend's rescue failed again, and I could only watch Decimal City recede into the distance, feeling utterly helpless.

The bird's talons shimmered in the sunlight. If the owl wanted to hurt me, it would have. The three other dots picked up by the owl seemed unharmed as well but just as cold.

The forest floor moved quickly beneath us. I caught glimpses of animals scampering in our direction. They were all working together. Other birds joined us in the air. A few large blue jays, owls, hawks, and even a couple of hummingbirds. All flying together, carrying dots in their talons and beaks.

The owl dove, causing a sudden weightlessness. Inches from the ground, the bird dropped us gently into the grass and landed a few yards away. It chirped and cooed with the other birds.

A few dots ran away instantly.

A group of us from Decimal City huddled together, too curious to flee.

The owl that took me landed, wings fully extended. A sudden burst of light consumed the animal, leaving 23 dots resting gently in the grass.

"Is everyone alright?" one of them asked, running towards the group.

Shocked, I stared in disbelief.

Animals of every kind were showing up all around us. First, the birds. Blue jays, parrots, woodpeckers, and hawks. Every animal carefully placed dots they retrieved from Decimal City and then, in a puff of light, transformed into multiple individual dots. Blue-jays became three dots. Hawks nineteen.

Woodland animals started scampering in from the forest. They would drop off dots and then shimmer into divided portions. Chipmunks became five or seven. Mice were three or five. A couple of gophers transformed into 11 dots each.

I stood in a sea of dots, cheering and celebrating after each transformation.

"Where is Lenter?" A concerned dot asked.

Head hung low, a dot that originated from a blue jay responded,

"They went full primal. There was nothing we could do. The forest claimed them."

Tears filled their eyes as they screamed grievances into the air.

In moments, thousands of dots covered the hilltop. The eagle I saw pecking at the top of the building swooped in swiftly and disassembled into about 73 dots.

The last animal to show up was the massive bear. Visible shallow cuts across its chest, arms, and legs. It laid down to rest. Exhausted from the fight.

This one didn't change. It whimpered and whined.

"Don't you go primal on me!" a dot from the eagle

shouted while in full sprint towards the beast.

"Come on, Adler! You got this. Listen to my voice!"

Now nearby, his steps slowed—as if approaching a wild animal.

The bear stood and growled loudly, teeth bared and claws visible. It stood on its hind legs, making it clear we were no match, and let out a deafening roar.

The dot from the eagle persisted. Staying low to the ground, they calmly approach.

"Come on, friend. Remember me. We are brothers. We have spent most of our life together."

The bear came back down to rest on all fours. It hesitated, groaning as if weighing whether to flee or to fight. Despite the size difference, the dot moved steadily and relentlessly forward.

With an enormous sigh, the bear gave in and collapsed to the ground.

Like a spring fully compressed, the dot circled the beast.

Only then did I see the blood gleaming on the bear's fur. They sustained serious injuries.

The dot from the eagle rested his tiny hand on the bear's muzzle.

"That's right." He said, petting the beast's forehead.

The bear whimpered in response.

The dot continued, "Now is not your time! You've still got a fight ahead of you."

Overwhelmed with emotion, his eyes filled with tears

as understanding dawned.

The bear's breath was getting shorter. It whined loudly.

He wrapped his arms around the bear's nose. Barely able to make it halfway around his muzzle.

"You did great!" he said, patting them on the head. "Rest now, friend."

The bear's eyes shifted from concern to peaceful. It drew in its final breath and let it out a massive sigh. Its chest stopped moving.

Tears filled my eyes.

Thousands of dots circled the beast. In unison, they let out a cataclysmic roar of a cry. They exchanged shouts of anger and sadness as if grieving. One by one, they approached the bear, placed a hand on its fur, muttered something between sobs, and moved on.

The dots from Decimal City remained together, observing.

They don't seem dangerous. I thought.

The dot that attempted to soothe the bear stood in front of us.

"Friends!" he said warmly. "I am Captain Numerous. We suffered a significant loss today, but it has bought you your freedom. We hope you will accept our sacrifice for you. We are Additioneers."

Cheers echoed around the field.

"The tragic nature of our work is that our numbers rise and fall." He glanced back at the bear as a tear fell from the corner of his left eye. He had a burly braided beard, incredible muscles, and a blueish gem dangling from his ear.

"We rescue dots like yourself from gnashers and Decimal City. If you are willing, we will gladly accept you into our ranks. Our goal is to free dots that cannot do so for themselves. You are free to do as you please."

A muscular, shirtless dot dropped a wooden chest next to him.

Captain Numerous lifted the lid and held up a few coins.

"With these amulets and the prime stones, we can shapeshift into animals. If you choose to join us, you will need your own paired with the Additioneers mark."

He motioned to the dot with no shirt. A tattoo rested across his chest. Spiraling specks outwards in a circle pattern.

"Should you join us, grab one and follow us back to our base, where we can give you a proper welcome."

He stepped back.

A plump-looking dot stumbled backward.

"I worked my whole life to get where I was!" He complained in a quavering voice. "You're ridiculous!"

He then collected himself and disappeared into the forest. A few others followed him.

A scruffy dot stepped forward. "Thank you!" she said.

Though torn and filthy, her makeshift clothing provided the necessary coverage. Her gaunt appearance suggested extreme hunger.

"Dots like the fat one," she gestured to the dot that had just run off, "are why I wanted to leave. It's built to benefit a few at the expense of everyone else. I never thought I would get out of that place." She bowed low and collected a coin. "I will serve alongside you."

The remaining dots reached the chest and collected a coin. None of them looked as bad as the first.

I was the only one left.

"And what say you?" The bear-chested dot asked. "Will you join us today?"

Confidently, I bowed and responded,

"I appreciate you trying to save me. I was there to find my lost friends. Sadly, I'm now separated from another friend. I need help to find them."

"A worthy cause!" Captain Numerous said in a deep voice. "We shall help you."

I felt relieved of a heavy burden. Much work remained, but the captain and his team should be able to assist.

"Delay your choice," he said, "but come with us. Tonight, we will rest and celebrate. Tomorrow, we find your friends!"

XIX

Additioneers

Opal

Travel in primes, they insisted. Smaller travel sizes are more agile and less likely to attract unwanted attention. If someone gets lost, it would be easier to identify—simply determine if the group is a multiple of another prime.

Sigrid led our group of 19 to the base of a towering cliff side.

"Our fortress is on top," Sigrid said, tossing some ropes to the ground. "I will tether us to one another, forcing us to move as one. Either we all succeed, or none of us do."

"Why can't we just transform into eagles and fly up?" a

dot complained.

She responded patiently, "The longer we stay in animal form, the more likely primal instincts will take over—we will become the animal permanently."

I pictured Wriggles blossoming into light and leaving a series of dots behind.

Sigrid tied the rope around our waists and continued,

"Prime means first. The first animals on the island were those created by the Sieve of Eratosthenes. An ancient relic from a culture obsessed with combining magic."

"What do you mean, first?" I asked. "There are eighteen numbers before nineteen. How can nineteen be a first?"

She shot me a glance of admiration.

"Well done! If you skip count by two's starting from two, there is no way to reach three. Three is the second first. Skip counting by three from three leaves four, but two can reach four. So it is not a first. Five is the third first."

"Oh, I see," I said, piecing it together. "Three is like a unique beginning. But there is no way to get to five skip counting by three or two."

"Exactly. The Sieve tribe eventually created the first animals. Believing animals were the purest form, they held rituals to merge dots into an animal form. This permanently united couples, families, and entire villages that loved each other. Dots are born of aether. Animals are born of firsts."

"You mean they can never return to their original form?" asked another.

Her cheerful expression turned somber.

"Another civilization grew out of malice towards the dots of Sieve. In their view, animals were unnatural and unholy. They spent their time dedicated to undoing this binding magic. They never fully developed the magic, but some horrific stories of their attempts. Their civilization remnants exist on the far eastern side of the island. Few dare to venture into their space. It's said a curse of their own making trapped them inside an eternal fractal."

The thought sent a shiver down my spine.

"Every dot ready?" She asked, tugging on the last rope to ensure it was secure.

We timidly replied. Newcomers were visibly shaking.

"You will find the climb easier than it looks. This rope is a precaution. We will walk most of the way."

She turned and started up the ledge. Each dot behind her followed her movements. About a fourth of the way up rested a sturdy path that allowed us to walk single file.

The next obstacle was a short ledge requiring us to hug the wall and carefully shift our weight to get across.

"Careful," the Additioneer ahead of me directed, "This is the hardest part of the climb. Let me know if you need a hand."

Before stepping on the thin slice of rock, I glanced over the edge to see how far down it was. This was a mistake. My head felt dizzy. I imagined myself falling. The thought of my body smashing into those low-hanging branches made me flinch. The dot behind me noted my folly.

"It's going to be alright. I will ensure you make it across safely." He said.

I froze. I remembered the falls I took trying to climb out of the hole Plex put us in. The dark caverns closed in around me.

That's when the guide placed his hand on mine.

"Come on," he encouraged. "I will be with you the whole way."

He instructed me where to place my hands and how to move across. I made it safely to the other side.

"Not so bad," a large dot boomed, clapping my back. The force knocked me off balance.

I attempted to push him back after finding my footing again.

He laughed heartedly.

The last step involved a jump to an unreachable rock. We untied ourselves before moving on.

I leaped, hoping it was enough to reach the other side.

Hands encircled mine as they hefted me to my feet.

"Welcome to the Fortress." A deep voice sang.

"It's good to be home."

Each of the Additioneers walks changed to a more relaxed form. With arms hanging loosely, they swayed, trying to shake off the weight of their shoulders.

An expansive field of grass and rocks rested at the top.

The fortress was circular, with archways for entrances and enormous gates to prevent unwanted entry. Symbols for North, South, East, and West rested atop each gate. Mostly built using large wooden trees, it provided a sturdy base. There were minimal gaps between the tied timber. A single tower rose from the center, with a platform on top where three or four dots stood as lookouts. The sharp architecture provided an intimidating atmosphere.

The inside was far more welcoming. An inner wall shorter than the outer wall provided housing for the inhabitants. The densely packed rooms could easily house hundreds of thousands of dots. Everyone was cheerful and welcoming. It reminded me of Countable Grove.

The delicious smell made my stomach rumble. A mix of tea, berries, and pastries. They led us to an interior space they called the town square, although its oval shape was somewhat confusing. The tower marked the center, and two platforms stood on either side of the focal points. There were no ceilings.

A room rested at the base of the central tower. Oversized squirrels, chipmunks, and a hedgehog were busy assisting with the chores. They grabbed berries, rocks, and even small branches.

The town square did not stay empty for long—dots of every size flooded in. Several dots went directly to the center, carrying hollowed-out pieces of wood. They blew, banged on, and plucked strings on the instruments.

Dancing erupted along with laughter and everyone

seemed to enjoy themselves.

A familiar, unpleasant smell, like sour berry juice, reached my nose. I sniffed my shirt to check if it was me. The scent evoked memories of the casino.

I'm celebrating while Mortimer is miserable.

The song ended.

"My fellow dots!" Captain Numerous said, standing on the furthest platform from me.

Despite the distance, I could hear him as if he stood next to me.

"Today marks a great tragedy. We lost 101 members."

As quickly as the dancing had begun, the room filled with tears.

He continued, "The battle was fierce. Some of our best gave their lives to ensure our success."

He paused for a moment, allowing his words to sink in.

"Death claimed 67 of the 101. The forest claimed twenty-three as owls and eleven more as a prairie dog. May our memories provide them guidance on their new journeys. May they find peace."

Thunderous whines of grievances rang through the crowd.

"Though they are no longer with us, they will remain in our hearts."

Cheers echoed around the room. A dot came walking by and handed me a wooden mug with a bluish liquid inside—the source of the sour berry smell.

"Their sacrifice was not in vain! Tonight, we celebrate

the success of saving 250 from the evils inside Decimal City."

A tumultuous roar filled the air.

"Enjoy the music and the wine!" He held a wooden mug high in the air, and every dot I saw followed suit. He drank.

There was a chaotic scene of music and dancing all around me. The songs were of the current and previous victories.

As hours passed, they placed candles and lanterns around the space to provide light. They lowered a tightly woven net across the fortress from the top of the tower. It must be to prevent rain from getting inside.

It felt wrong to have fun, knowing Mortimer might be in pain.

The next day, I woke in an uncomfortable position. The town square was littered with sleeping dots. A few of them had woken up and looked as confused as me.

"Young one!" the captain said jovially. "It's great to see you enjoyed yourself. Come with me."

I followed him to a room at the base of the tower. A few other dots were inside, analyzing a miniature model of Decimal City.

Captain Numerous pointed to each dot in the room and introduced them.

"This is Dag, our strategist."

Dag was busy looking at the city model, clearly deep in thought.

"You already know Sigrid, our fiercest fighter."

"Nice to see you again, Opal." She said with a bow.

She didn't look like a fierce fighter at all. Average size, build, and friendly.

The air around the fortress trembled with the sound of a long, vibrating horn.

"Intruders to the west!" a voice echoed from the tower above.

I followed as the captain and others prepared for a fight.

Sigrid picked up a quiver of arrows and a bow. She was the fastest of the group. Within moments, she stood atop the west gate and looked beyond.

"Eight dots in total!" She shouted back. "A pair walking from the southern side and six walking due east."

"Hold your fire," Captain responded. "I will go out to meet them!"

Dag and Leif were at his side, and I watched from the west gate.

As the intruders stepped into the open field, they sharpened into focus. My heart filled with delight. I burst down the ladder.

"Stay here!" Sigrid said, "It's not safe!"

"They are my friends!" I shouted back. I ran out of the gate and into the open field.

"Stop!" I yelled. Waving my arms, "They are with me!"

With each step, tears fought their way to the surface.

They are alive. I thought.

"You know these dots?" The captain questioned.

My smile said it all. His face melted from a look of concern to a welcoming grin.

"There are two others. Do you know them as well?"

A lump swelled in my throat, preventing me from speaking. I shook my head no.

"Very well. Take Lief and Dag with you and greet your friends. I will meet up with the other two."

My vision became blurry from tears. Without hesitation, I continued to run.

"Opal?" Ellis shouted.

The Caretaker stared in amazement, a mixture of shock and wonder on her face.

Ronan was speechless. There were three other dots I didn't know, but I was pleased to see the familiar faces.

Ellis opened her arms, and I dove in. She wrapped me in a satisfying hug.

"What are you doing here?" The Caretaker asked, wrapping her arms around Ellis and me.

"It's good to see you," Ronan said.

For a moment, I melted back to my first few days in Countable Grove. Their warm, smiling faces looked down at me.

They gave me my name. I thought. *The name that didn't feel like mine until I met Tia.* It felt like a distant memory, but at this moment, it was closer and had a new perspective.

I'm proud of my name. I'm glad they gave it to me.

"I thought you were dead." The words burst out of my mouth awkwardly.

"Not dead," Ronan said. "Just captured and imprisoned." He gestured to two dots that had vines on their face like the officers in Decimal City. "By these two."

I caught his sarcasm instantly.

"Allow me to introduce you to Dux and Phi." He said, pleased with the discomfort.

He indicated two figures with faces obscured by vines. Neither of them seemed comfortable after his introduction.

"This is Oizys," Ellis said, pointing to the final dot.

He gave a bow, and we started back towards the fortress. Our paths merged with Captain Numerous, a dot named Lyra, and another named Quint. Reunited, we walked into the fortress.

XX

The Square Root of Two

Eulalia

All of us were glad the caverns were behind us. I found a new appreciation for wide open spaces and plenty of light.

Ronan appeared exceptionally joyful to be living. Each time we paused, he would interact with any living thing he could. Flowers, grass, trees, and even an angry squirrel. After taking some food, the squirrel became friendly. It would run up and down his leg, and eventually it rested on his chest for an afternoon nap.

Ellis hadn't left Ronan's side since he came back.

I felt relieved the tension between all of us had

dissipated. They seem back to their usual selves.

The three of us, Opal and the three leaders of The Negative Caverns sat in a room with a model of Decimal City.

I placed the square artifact on the table.

"This is what we recovered from the Crypt," I said.

Captain Numerous stared at the artifact in deep thought but did not touch it.

Dux bounced nervously in his chair.

"I see the beans still curse you." The captain said, turning to Dux. "Have you held the artifact?"

"I'm afraid," Dux said in a shaky voice. "The last time I touched something they created, I ended up like this."

He gave a nod of acknowledgment.

"Were there any other secrets you uncovered in their crypt?"

"Just one," I responded. "Their lies cursed a Babylonian tablet. The curse nearly trapped Oizys as a statue. Uncovering the truth that the Pythagorean triples are Babylonian triples freed him. Ellis still holds the artifact that helped undo this magic."

"How did the magic work? Did Oizys need to touch the artifact?"

"We just needed to bring it close to him."

He let out a deep sigh and studied the artifact from multiple angles.

The room waited in quiet anticipation.

"I think a member of the brotherhood needs to offer

you the artifact." He said with a nod. "I feel there's more to this artifact than meets the eye. A curse will befall any brotherhood member who gives the artifact to Dux, lasting until the discovery of their final secret."

Welcome to the brotherhood. The ominous words engraved above the artifact flowed into my mind. *That makes me a member.*

"I volunteer," I said, grabbing the artifact off the table. "I will carry this burden with you until we uncover their secret."

"Very well," Dux said.

A new tension bound the room.

Phi nervously awaited the results.

I offered the artifact to Dux. The vibrational energy flowed across my whole body.

He stretched out a trembling arm.

The gold diagonal glowed a warm orange-yellow. A mysterious wind blew maps around the room. Light filled the space.

I shielded my eyes from the brightness.

Dux and I floated off the ground. As Dux made contact, an explosive energy wave flew out from the artifact.

The force of the explosion destroyed the walls and sent everything flying away from Dux, me, and the artifact.

I stood in a dimly lit grass field. A wooden fence marked

the perimeter. A dot, about one-third my size, was playing with a ball in the field. They kicked the ball against a fence post so precisely that it bounced back to them.

"Come play with me," they giggled.

The ball rolled towards me and stopped just in front of my right foot. It shifted between 3D shapes. One moment, it was a jagged Tetrahedron. Next, it morphed into a sphere—A kickable surface. Before it could shift to a different shape, I booted it back. As the ball flew, I heard a voice.

"Help me!" it pleaded.

I looked around, but the fog was too heavy to see anything past the fence.

The small dot laughed as they bounced the ball off their head between shapes.

I ignored the voice and kept playing with them. The game required us to kick when it was most round. If we kicked it when it was a cube, tetrahedron, or octahedron, it would turn red, become heavy, and fall to the ground.

"Please, I need help!" Dux's voice cried out.

"Did you hear that?" I asked in a panicked tone.

The ball fell to the ground.

"Hear what?" snickered the dot. "Are you hearing things?"

My eyes darted around the grass field. I took a step forward.

The ball rested in front of me again.

"My name is Even!" the dot chirped.

I smiled and kicked the ball back to him when it was an icosahedron.

"Eulalia," I replied, trying to mirror their cheerful tone.

Reaching the perimeter, I gazed out at the dense fog.

"Help!" Dux screamed.

Sweat crawled down my forehead. I jumped the fence and ran into the fog towards the sound.

"Stay close to me," I yelled back, waving to Even. "I'll keep you safe."

He ran towards me with the ball in his hand. It morphed back into a cube shape.

The cube reminded me of a memory—something I had lost. It was something important, but it was difficult to place its significance.

My heart pounded in my chest. We ventured into the fog, losing sight of the fence behind us.

"Do you know where the field went?" I asked Even.

"There's no field." He giggled.

In one motion, he absorbed the ball into his chest and ran into the fog.

"Wait!" I shouted. "Stay with me!"

It was too late. The fog washed away to a dark gray. The wind blew my clothes and hair.

"Help me! Please help me!" Dux's voice rang out again.

"Where are you?" I screamed. "How can I get to you?"

I ran towards the voice.

"Come now, please!"

But now it was coming from the opposite direction.

I stopped. Another dot rested in front of me—or maybe an animal?

They had a sibling resemblance to Even but had fur and claws.

The creature was crying.

I reached out my hand to comfort them.

They turned, showing their teeth, hissing and growling.

I backed away.

"Why didn't you play with me?" The creature growled.

I tripped over the morphing ball behind me. This time, it had a dark energy flowing from it.

"I didn't know you wanted to play. Here," I kicked the ball towards the creature.

It caught it in its teeth.

Dux's voice rang out gain.

I turned to the source. I could see him.

Dux also had a sibling resemblance to both Even and Odd, but the morphing ball was his head.

"Play with me!" the creature demanded.

They ran towards me, growling loudly.

I dodged to the side.

"Even, is that you?" I asked.

"My name is Odd," they snarled.

I dove again. His claws narrowly missed my face.

The events were a blur as my mind struggled to piece everything together.

Dux looked like Even and Odd.

How could he be both? I thought.

I dodged the creature again and kicked the ball towards them.

"Let's play," I tried.

The ball slammed into its face, absorbed into its body, and morphed back into the dot named Even.

"Please don't hurt me!" He said, tearing up. "I just want to play with you."

"This doesn't make any sense!" I yelled in frustration.

"Where is the ball?" I tried.

"It's just there."

Floating in the air behind me was the metal artifact with the gold diagonal. Glowing brightly. In an instant, the events clicked into place.

It's not supposed to make sense. That's the point. If Dux were rational, he would be Even and Odd. That's impossible.

"It's a contradiction!" I yelled. "Proof that the diagonal of the square is irrational. None of this is real. That's the brotherhood's secret. They found proof that irrationals exist and hid it away. They lied about their discovery!"

The gray faded to the warm dirt of the fortress. The artifact released an explosion of energy.

I found myself on the ground next to Dux, the artifact between us.

His face morphed and shifted in unpredictable patterns—no longer bound by the vines.

"You're free!" I cheered, jumping to my feet.

All around us were shards of boards. The explosion destroyed every part of the strategic room except for the towering post. The ladder, however, only came down about three-fourths of the way.

"Are you ok?" Opal asked, leading the group of dots back to us.

I gave a nod of encouragement.

"You did it!" Phi said, observing Dux.

Vines still plagued Phi's face.

Dux was now standing majestically behind me.

"Thank you!" He said. "I remember now. I am The Square Root of Two. My efforts and stupidity caused this plague to happen. There were signs of their treacherous plans. They promised they would never. I was too naive."

Quint rushed toward the artifact lying on the ground and picked it up.

Nothing happened.

She let out a shrill noise of pain.

"Why doesn't it work for me?" She demanded in desperation, clawing at the vines on her face.

Phi was immediately at her side, attempting to restrain her arms.

The captain said, "I think that is because that secret is only for one specific irrational. I do not think the unbinding magic will work on any others. Dux is the first to be bound by this curse. Therefore, proving his existence as the square root of two freed him. It provided the truth to the lie."

Quint laughed maniacally.

"Clearly, I exist. I am standing right here."

"I know you exist," He said, trying to reassure her. "If I were the aether that provided this world with magic, I would free you. Alas, I am not. Our creators require further convincing."

Quint fell to her knees, screaming. She scrapped at the vines on her face, turning the scars red.

Her desperation pained my heart. Trapped in a foreign body, unable to be her true self.

Dux placed a hand on her back.

"My memory is coming back." He said. "Before I found the Pythagorean Brotherhood, I spent a moment with a rather curious tribe of dots. The dots of Al-Khwarizmi. They fought against Gaia's controlling nature. They aimed to form a natural order that managed itself and required no interaction from a higher power. I think their work could prove useful and may provide a cure."

Quint's convulsive sobs slowed.

"I'm in," She said through gritted, angry teeth.

Her ferocity scared me. I could see an emptiness inside her—something begging to be released.

"I am, too," Opal responded.

The dots in the room stood up.

Oizys stepped forward.

"Whatever it takes." He said.

Caution jolted inside me, "We need to prepare for the cursed dragon." My reminder shocked both Captain Numerous and Dux.

Dux stared at me in shock. "Why?" He asked.

"Arithma intends to wake the beast and may have already done so."

"Surely she knows that would be a mistake." The captain said. "The cursed dragon will devour the whole of the island."

"Phi captured us the day we found out. We are trying to unite the five domains through a council."

"It appears," Dux responded. "By a stroke of luck, that we share the same goal, but two different names. Al-Khwarizmi's work was a compilation of the known power of their day. A raw form of magic, pure and built from the roots of the island. As potent and simple as counting and as powerful as the mountains surrounding us. They called this work The Council of Arithmetic."

I felt a surge of power telling me it's true—something calling from the depths.

"Five domains across the land united under one magical order. The order requires a councilor from each of the five domains paired with a life force artifact."

Natural order? I thought—*the way I allowed Countable Grove just to grow. I stepped back as a leader. The dots inside took care of each other. Could this be the answer I've been looking for?*

I glanced around the room, making note of which domains were present. Dux, Phi, and Quint were from the irrational domain. Any dot from Countable Grove was natural. Lyra was the only dot from the negative domain. We were only missing servants from two, the rational and the complex domains.

"How will we find a council member from the complex domain?" Ellis asked.

"A fair concern," Dux responded. "That domain is volatile and challenging for most. It will be difficult to find a willing councilor."

"Has anyone seen a dot from that domain?" I asked.

Opal excitedly raised a hand. "Mortimer and Plex! They helped me escape the gnasher."

"Wonderful!" Dux said, "Where is he now?"

"He's in Decimal City." She responded sheepishly. "They captured him the day you rescued me."

"We will help you save your friend." The captain responded in a confident tone. "That leaves only the rational domain. I suppose any partialed dot would do. But we need an artifact to represent the life force of the domain. That brings us to the mountains near Countable Grove. The core of the mountain provides a pulse. The trouble is, that Nymphaea is difficult to persuade and very dangerous. Whoever ventures this way is not likely to make it out alive."

Dux responded, "Who will go to retrieve the fruit on top of the mountain?"

Ronan and Ellis raised their hands.

My heart sank.

Ronan said, "If they are close to Countable Grove, I want to understand the threat."

"I will travel with them," I said, unwilling to risk losing them again.

"Very well," Dux acknowledged, "all remaining dots will focus their efforts on retrieving Mortimer from Decimal City. We will meet up at the cave entrance north of the city."

XXI

The Prime Ceremony

Opal

The warm sunrise colored the sky at the east gate. We watched in anticipation of the prime ceremony. Last night, we inaugurated new Additioneers. The inauguration poses challenges that the newcomers must overcome to become fully accepted as Additioneers. From what I understand, it's all ceremonial and wouldn't *actually* prevent anyone from joining their ranks. However, today is their official prime ceremony—their first day as an animal. They will carry Ellis, Ronan, and The Caretaker to Nymphaea's mountain.

Captain Numerous ran to join us.

The recruits' tattoos looked painfully red, but showed signs of healing. Though they did not choose the plus sign centered in a circle rune, they decided where to place it. Some were large and noticeable, others subtle or hidden under clothing.

The Caretaker leaned over to me.

"You are proving to be quite resourceful." She praised. "You have accomplished a lot in a rather short amount of time."

Me? I thought. I felt my cheeks grow hot as embarrassment flushed my face. The thought I was doing a good job never crossed my mind. I was losing friends with every turn. I met her gaze. Tender, kind, and heartfelt. *She meant it.*

"Thanks," I mustered in response.

Captain Numerous was now by our side.

"I always love watching these," He whispered, joining us outside the circle.

The group of 23 dots jumped in rhythm.

Bounce-bounce-jump.

Initially, their heads were uneven as the newly inaugurated dots tried to keep up. Slowly, they moved as one. They introduced a new piece of the dance, one at a time. The leader, Sigrid, spun in 360 degrees, maintaining the jumping pattern.

Captain Numerous gave us insights into the purpose of the dance. "Moving together as one requires significant synchronization. This part of the merging process ensures they can work together once they become the animal. Every dot is reliant on the leader to guide them. They need to feel the animal but avoid giving in to their primal instincts. Each of them will have a role to play as they work together. Smaller animals are easier to maneuver as they require fewer dots to form. The larger the animal, the easier it is to lose sight and give into primal instinct."

The group of dots chanted. They sounded like murmurs and whines. Almost like a happy form of crying. They continued dancing in circles as each slowly matched Sigrid's movements. All of them were moving as one, anticipating the next step perfectly.

Sigrid held up a gemstone that glowed brightly.

The remaining 22 members each held up their golden prime talismans. The prime stone emitted a rhythmic pattern as if communicating, and talismans lit up in reaction. A pleasing vibration filled the air. A series of notes that resonated well together. As they danced to the pulsating music, they moved closer together.

Standing with their shoulders aligned, their bodies blended together. With each phase, different sections of the animal became more visible. Feathers erupted from the center. The head and beak shot upwards. In one fluid motion, they sprang from the ground, flying upwards.

The sight of the owl circling the fortress was breathtaking.

"Your ride is here," Captain Numerous said, breaking the silence. "They will take you to the mountains just outside Countable Grove and wait at the base for your return. Be cautious around the mountains. Nymphaea is a ruthless leader. If you get caught, she won't show mercy. Travel through the mountain, as it will save you travel time. Sigrid will then bring you to the entrance of Al-Khwarizmi's study. Fly safely. You are always welcome to our fortress."

He bowed and stepped away.

Before I could follow, Ellis pulled me into a hug.

"You are shaping up to be quite the witch! I'm growing quite fond of you." She said with a warm smile. "Take care of yourself. I hope to see you again in Countable Grove."

Once Ellis released me, Ronan wrapped his arms around me. "You are full of surprises, little one. I'm glad you got trapped in that cave." His face went red with embarrassment. "Um... I mean. Sorry. That was stupid." He was fumbling over his words. "It's good you were in that cave, is what I meant."

He slapped his forehead.

Ellis laughed. "I think he is trying to say we are glad you are with us."

"Yeah, right... That." His eyes darted towards the ground.

"Thanks!" He said.

The three stepped away just as the owl swooped down and picked them up. I watched as they shrunk into the distance. A warm peace rested in my belly. It reminded me of how I felt with Tia. Like my existence was all that was necessary. I smiled, overcome with the realization I was no longer alone. Tia, Aleph, and Mortimer were being going to be saved by all of us working together.

"Come with me. We need to make preparations." Captain Numerous pointed back towards the fortress.

I complied.

The platform and shed that once marked the center were still a shattered mess. A few tables were pulled together, with a couple of maps strung out across. Four dots stood around the table.

Oizys looked restless as he studied the map of Decimal City.

The strategist Dag stood beside him, pointing out a few places on the map.

"Maybe they are here?" Dag said.

"Out in the open? That seems like a strange place to keep them," Oizys responded.

A few dots stood just out of earshot, obviously trying to listen. They turned abruptly once they saw us.

Leif approached the tables from the other side with a few dots I didn't recognize.

"Captain!" He said with a bow. "This is Craig and Whist. Both of them are knowledgeable about the inner workings of Decimal City. They have agreed to help free Mortimer and Plex."

"Wonderful!" Captain Numerous responded. "Have a seat, will you."

Someone pulled up a few chairs, and I sat across from

Oizys, who remained standing while gazing at the map.

"Let's start with the most obvious question. Do you know where they take dots with enormous debts?" Dag asked.

Both of them looked towards the ground.

"Unfortunately, no," Craig said. "I only heard rumors most of the time I was there."

The dot named Whist remained silent. She looked nervous. Her right hand gripped her left elbow against her body. She avoided eye contact.

"And what about you, Whist," Dag said, lowering himself to ensure he made eye contact.

She was shocked. Looking around the room, she saw how many people were listening.

"It's alright." Captain Numerous responded. "This is a safe space. Take your time, and we will wait as long as necessary."

Dag shot him a look of surprise and mouthed, "As long as necessary?"

At a certain point, he considered disputing or challenging the idea. Recognizing the pointlessness of his actions, he gave up and sat heavily in the chair behind him. He let out a frustrated sigh and took a special interest in the grain details on his chair.

We sat in silence as Whist turned bright red. She swallowed, took a deep breath, and said, "They take them to the back of the tallest building. I don't know what floor, though."

"I knew it!" Dag said, jumping to his feet. He placed his hand on the map. "Look right here. In the open, like I said!"

"Thank you!" Captain Numerous responded to Whist. He gave a disapproving glance towards Dag, then asked, "Do you know how to get there?"

Her reaction showed that she was not expecting any follow-up questions. She shook her head no and looked at the floor once again.

"It's at least a lead," Oizys said. "I don't think they would store their prisoners in the open like that."

"Why not?" Dag retorted. "Everyone in Decimal City is a prisoner. They are all out in the open."

"You make a fair point," Oizys said. "Think we could simply show up and find it?"

"Very well. We should only have a few of us inside. Then, when we need to escape, we should have strong numbers outside ready to fight." Dag recommended.

Captain Numerous said, "Very well. Oizys and Opal will come with me into the city. We will notify you when we need to escape. Dux, your powers will prove useful to provide a distraction as we escape. Are you willing to help?"

"Certainly, this will help us free my siblings," He said with a bow.

I tried not to stare, but his head continued to shift between shapes in unpredictable patterns.

Captain numerous said, "We will take enough dots with us to ensure we can escape. Dux working with us should provide a significant surprise. Dag, I need you to lead the charge as I will be inside the city."

"Very well," He said with a bow.

We took the rest of the afternoon to make preparations. It was hard to sleep that night. I had conversations with myself, wondering what horrible things could befall Mortimer.

Hundreds of dots awaited departure. Noticing who I was, they stepped aside, providing a clear path to Oizys, who took a calm drink of water.

"Nice day for a jailbreak," he said calmly, eyes closed.

I sat down next to him. I could feel the warmth of his shoulder radiate against mine. It made me uncomfortably aware of how chilled I felt.

"I suppose it is," I said in response.

Captain Numerous joined us within a few moments, causing a chain reaction of dancing and chants. Groups of dots were undergoing the prime ceremony. Each one provided a different chant and a unique set of jumps. The space felt chaotic, but focusing on just one group at a time provided a harmonious chorus.

A group of three were bobbing up and down in complete unison. Their circle was getting smaller as they transformed into a field mouse and scampered off. Multiple groups of thirty-seven transformed into wolves and ran off in a pack. I noted a few prairie dogs and a fox.

A brown owl swooped down and picked up Oizys and myself, followed by a midnight black owl, which picked up Dux, Phi, and the captain.

The frosty morning air sent a chill down my spine. I wanted to land quickly. It was too cold to fly. This was the second time I had traveled by an owl, and it was unpleasant. I rubbed my arms with teeth clenched as I watched the sunrise over the fog of the unknown.

Sunlight broke through the clouds, separating the orange, pink, and purple hues. I wished seeing something so beautiful did not require such an inhospitable environment. Within a few moments, we plummeted towards the ground.

I found myself placed carefully in a forest just outside Decimal City.

Oizys threw his cloak over me for warmth as I shivered in place.

"Thank you," I said between chattering teeth.

The brown owl transformed back into 23 dots. The black owl dropped Phi, Dux, and Captain Numerous, then fluttered off.

"Decimal city," Phi said.

The large numbers in front of the wall were visible from our vantage point. 4 2 6 2 3 4 | 4 9 8. I couldn't remember the numbers from the last time I visited. However, I was confident the Additioneers had their work cut out. It would take a great deal of effort to free them all.

"We will monitor things out here. If we see anything

suspicious, we will move to get you out. Good luck!" Dux said.

I forgot just how large the walls outside the city were. The line to get inside was more varied than the last time. We watched patiently as the numbers increased, showing more dots trapped inside. We soon stood in front of the admittance officer.

He wore a helmet over his head and shiny metal armor covered his body.

"How many?" the guard demanded.

"Three," I said confidently.

"Very well. Move forward." They motioned for us to walk through the gate and shouted, "Three more!"

I took a deep breath before walking through the entrance.

XXII

The Rational Domain

Eulalia

As we gained altitude, my troubles remained grounded. With each breath, the chilled air renewed my energy and froze my anxiety.

It's beautiful. I thought, looking over the island. The owl held me securely but gently in their claws. I let out an enthusiastic call.

"I always want to travel by bird!" Ellis shouted, arms extended.

Ronan clung to the bird's claws.

"Just let me know when it's over," He yelled.

With a smile and a chuckle, I surveyed the land. Trees

were just smudges from this height. Surrounded by sea, the Countable Forest was much larger than I ever considered. Despite my adventures through the sturdy trees, this vantage provided a unique, almost foreign, perspective. I couldn't see Countable Grove. The mountain we were approaching blocked it from view.

I hope they are safe. I thought.

Decimal City was impossible to overlook. Roughly a quarter the size of a Countable Forest. Everything was smooth and crisp. No birds overhead or animals playing.

I felt a ping of pain in my heart, considering what life would be like trapped inside. No funny squirrels to watch or flowers to smell, just walls all around.

The owl placed us on a ledge near the top of the mountain. We watched as the bird made a couple of passes before diving towards the base.

Ronan was pleased to be on solid ground. He collapsed to the ground and let out a sigh of relief.

The ledge was just large enough for the three to sit with a life-threatening drop off on either side.

"I guess we are on our own from here?" Ellis said peering over the edge.

"Looks like it," I responded.

"Anyone know what this fruit looks like?" Ronan asked.

I shrugged. "They said it was on top of the mountain. Hopefully, it's clear when we get there."

I turned to look at the wall standing between us and the summit. I tested my weight on one of the nearest footholds. There were enough protrusions to make climbing the rest of the way arguably safe. I climbed up first.

A vast granite engraving greeted us at the top.

May multiples provide you with light
That ye may always find the lair
Lest ye become consumed by dark
Travel in pairs

Though the surrounding ridge resembled a dormant volcano, the interior was surprisingly flat. In the center was a tranquil pond with eight short, stubby trees evenly spaced around its edge. The trees were about my height, with delicate pink leaves.

"What a weird greeting," Ronan said, tilting his head and trying to imagine what it meant.

"Multiples are the numbers that occur from skip counting, right?" Ellis asked.

"I think so," I responded. "The numbers two, four, and six are all multiples of two. Three, six, nine, and twelve are all multiples of three."

"This feels eerily like the crypt," Ellis said.

A deafening silence fell over the three of us.

Why couldn't the owl just land us near the pond? Though there was no immediate danger, I couldn't help but feel things were not as they appeared.

"They said they would meet us at the bottom?" Ronan

said, peering over the edge. "That climb will take us days."

Ellis studied the rock and its message, bouncing with nervous anticipation. Realizing it wouldn't give us anything further, she took a step beyond it. The mountain rumbled as the ground beyond fell away. Rock platforms wound around the pond.

Ellis stepped back to this side of the engraving. The mountain shook until the ground appeared solid again.

"Those platforms looked like the way to get across." She observed.

"I want to get a better look," I said, stepping forward.

The ground fell away, as before, leaving a set of platforms close enough to leap between. The solid surfaces were just large enough for one dot at a time.

"I think it's skip-counting steps," I said. "It's providing a platform for every other step."

"I think you're right!" Ronan said, stepping forward to join me.

The earth trembled, and a new expanse of solid platforms emerged to the right. Spaced every third, made it too far for me to jump.

Ellis stepped forward. An additional set of platforms joined the others. These were five steps apart.

"Travel in pairs," I said, feeling disappointed. I came to protect them, but may need to sit this one out.

"We can't just leave you," Ronan said.

"I'll be fine. I've spent plenty of time on my own. Go retrieve the fruit so we can be on our way."

Both Ellis and Ronan eventually gave in to the logic.

Their movements between platforms were fluid, requiring only brief separations every three jumps.

Ellis took the path of every other step. She provided a sturdy foundation for Ronan to jump across the more significant gaps.

They disappeared behind a few of the trees for a moment.

"We made it!" Ronan shouted.

I cheered.

Ronan carefully grabbed hold of some fruit and plucked it off.

The mountain shook violently. The platforms fell into the darkness below.

A wave of dirt shot out from the trees of the pond, and the floor was solid once again.

I took a timid step, but it remained sturdy. I tested my weight on the part that initially fell away. Satisfied that it wouldn't give out, I walked cautiously forward.

"I think we can step anywhere now," I said.

Ellis grabbed the fruit as well.

"Now what?" Ronan asked. "Climbing down will take days."

I was now in the center as well and plucked another fruit. They smelled like berries mixed with citrus.

"I don't see any immediate danger," Ellis said, as she sat down to rest under the pink leaves.

I shrugged and joined her.

Ronan remained standing.

"That was too easy," he said, remaining alert.

"We are not alone," Ellis said, standing up quickly.

At the base of each tree, a tiny, nearly undetectable village of dots sat.

The three of us stared in disbelief. Upon closer inspection, we came to recognize the deceptive nature of its size. The village was rather large. There were thousands of dots going about their day in a space no larger than my torso.

"Intruders!" a squeaky voice rang out. High-pitched bells tolled from all the trees.

In an instant, the ant-sized dots became twice our size and surrounded us. In a single fluid motion, they threw us to the ground and gagged us as they bound our hands and feet. The amulets around their wrists glowed brightly. Trees grew taller, and their leaves became boats. The grass, once no taller than my boots, became a forest, and the dirt below us was like boulders. We shrunk down until the tiny colony of dots was as large as Decimal City itself.

If Nymphaea captures you, she will not spare you. The captain's words rang in my mind. My muscles tensed as I fought back against the bindings. They were too tight. The clump of fabric, which tasted like dirt, became tighter as I struggled. It was hopeless.

The dots carrying us were bulky. Their muscles rippled through their clothing. Their heads resembled a drop of water and their armor flowed down their body, highlighting their cut and jagged figure. A roaring crowd filled the air as the dots carried us to our captor.

"Infidels!" a dot screamed in my ear.

The guard effortlessly kicked them away and continued onward.

The ever-increasing number of observing dots poked and prodded my ribs with long rods. Each jab sent a shock of pain up my spine. The guards threw us to the ground in front of their leader.

The obstreperous crowd behind a chained fence chanted grotesque victories of our death.

"Kill them all!" voices screeched.

A dot that was twice our size rested on a throne. They wore a large crown with robes draped across their body lazily. Despite their calm appearance, they bore scars across their chests and arms. A large scar down the right side obscured their face. They seemed inconvenienced and disgusted by our presence.

The dot on the throne did not address us. Instead, a dot that looked out of place stepped forward. They wore a sparkling purple gown that flowed down their body like water. Moving droplets highlighted the edges of the heavy fabric. A crisp contrast to the grungy, dirty appearance of the dots surrounding us. Each step sang grace and elegance.

The room grew quiet as she approached.

"What have you brought before us today?" Her voice was deep and powerful.

"Intruders!" the guard squeaked with a bow. His hands shook.

"They stole our fruit but did not eat. Such a thing is punishable by death!" another guard snarled. The grungy looking observers broke out into cheers.

"We want justice!" A wispy voice shouted from the crowd.

The elegant dot lowered herself to my level.

I felt her icy fingers take hold of my cheeks. A stab of pain shot through my jaw forcing me to open my mouth. I stared into her cold, lifeless eyes. She took special note of my right eye and cheek.

"They seem healthy," she said, inspecting my open mouth.

Her deep blue fractal eyes pulled me in. They created a never-ending growing sensation. It was mesmerizing until the shot of green interrupted the movement.

"A primitive group," she declared, pushing me to the side and noting the artifacts in my coat. "They cling to old ways! They are not from here!"

A hush fell over the crowd.

"But, how?" A trembling voice asked. "We weren't supposed to be visited by outsiders for another hundred years."

A deep, crooked smile grew across her face, eyes ready for blood.

"It appears our seers have lost touch." She said. Her eyes locked in with mine. "A severe punishment awaits them.

But for now, leave me with them. I need to speak with our intruders *alone*."

Within moments, the chamber emptied, including the dot from the throne.

The elegant dot walked to a table with a pitcher. They poured the contents into a cup and drank some.

"I am Nymphaea," she said. "Give no heed to the chants of the others. They are harmless."

She placed the cup on the table and walked over to me, and pulled the gag out of my mouth.

"Now, what are you doing here?" she asked, eyes cold and dead. The mesmerizing pattern of blue fractals was now dark and empty.

"We mean no harm," I tried. "We need some fruit to save some of our friends. Please let us–"

She pulled the gag back over my mouth.

"I've heard quite enough, thanks. *You poor things.* You just want to help your friends." She walked over to Ronan and pulled down his gag. "What about you? What are you doing here?"

"We have what we need. Let us go."

She made a disappointed noise and put the gag back on.

"No. I enjoy having you here. You three could be my new pets."

She circled Ellis.

"This one looks the least fortunate of the group. What is she doing here?"

She pulled off her gag.

Ellis spat in Nymphaea's face.

"Let us go!" She demanded.

Nymphaea didn't flinch. She didn't even try to wipe it off.

"This one has a bit of fight in them. What a pleasant surprise."

She tossed the gag to the ground and sat on the throne, tapping a finger to the corner of her mouth as if looking for something.

"I had everything perfect until you three stumbled in like blind rats," Nymphaea said. "Unfortunately, you are about one century early."

"For what?" Ellis asked shortly.

Nymphaea let out a cackled laugh.

"Goodness me, no need for such tones. I'm getting there." She took another drink from the glass she was holding. "You were going to bring justice as demons from beyond. However, you are not daemons, are you?"

"You're sick!" Ellis spat, trying to wedge her arm out of her restraints.

"I've known for a long time there are no daemons from beyond." Nymphaea let out a reminiscing breath. "I've needed to maintain an image. It's a troublesome burden to carry. To be the one ordering dots to burn homes and slaughter thousands to keep up the ruse."

She smiled.

"Why?" she asked, "So I can claim them as the very

daemons from beyond. So, imagine my surprise when you three show up here clearly not to murder and kill. I can't have those types of beliefs among my people. They would fight back. Where would I be if I allowed that to happen? Certainly not here."

She grabbed a large knife resting against the inside of the throne. "I will need to kill you before rumors can spread. How unfortunate for you."

Ellis stared, speechless.

We were bound and unable to cast magic. Frantically, the three of us shifted from side to side, trying to break loose. The knife she held created a stark contrast between her demeanor and her plan. Her eyes held an undeniable, insane gleam.

From the darkness behind us, an explosive noise filled the air.

The doors blasted open.

"Hypatia! Cut them free and get out," a dot shouted while jumping over us, deflecting the knife in Nymphaea's hand and knocking her back onto the throne.

The ropes binding my wrists slackened and fell to the ground. I pulled the gag out of my mouth.

As I turned to thank my rescuer—nothing was there.

Rather than stand and argue, Ellis, Ronan, and I sprinted out the doors. We dashed down the corridors, trying to find an exit. We could hear guards on the approach.

"This way," a voice shouted from a staircase to the right.

We sprinted down the stairs and deeper into the mountain. After putting some distance between us, we rested. We ate some berries to replenish our energy.

"Please keep moving," the voice encouraged.

"Where are you?" I asked.

"Who are you?" Ellis corrected.

"I will tell you at the base of the stairs. Please hurry, there is no time."

I looked over the edge of the stairs to the complete darkness below. With no alternatives, we pressed on.

An archway greeted us at the bottom of the stairs. On the other side was a cave. Pulses of orange and red light radiated across the rocks.

Something fell off my shoulder. As it flowed to the ground, it grew into a dot until it was about our size. She had a water droplet for a head. Her clothes, faded and patched, hung loosely on her thin frame.

"I am Hypatia, a member of the inverse order. We've known for some time there were dots from beyond like yourself, but Nymphaea kills anyone that suggests otherwise. It doesn't help that she has brainless buffoons following her every order."

"Thanks," I muttered between breaths.

"We need to keep you alive," she responded. "The prophecy says you will destroy our entire civilization. Your very existence provides proof that may help us turn the tides of this never-ending war."

"Glad we can help," Ronan said, straightening up.

"I can show you the way out," Hypatia responded. "Did you get what you needed?"

"Yes. We–" I checked the pocket where I placed the fruit—empty.

Ronan's and Ellis's fruit were gone as well.

"No. I suppose we don't," I said in frustration.

"Is it just the fruit of the tree that you need?" She asked.

"Yes, but ideally, it would be the source of the fruit. The gemstone provides the fruit with life."

Hypatia spun around in shock.

"You want our life force?" she asked, face staring in disbelief.

"Well, at this point, I just want to get out of here alive," Ronan said. "If we do so without the gemstone, we will need an alternative."

"Taking our life force would destroy us all. It's what provides the entire mountain with life. What do you intend to do with the gemstone?" She asked, nervously bouncing between her feet.

"Are you familiar with the tale of The Cursed Dragon?" I asked.

"Never heard of it."

"It's a beast that will devour all. We need the gemstone to fight back against it."

"We didn't know your situation," Ellis responded. "I'm sure we can find another alternative."

Hypatia's arms relaxed by her side. The words seemed to offer a bit of comfort.

"Do you have any fruit from the trees atop the mountain?" Ronan asked.

"Unfortunately, no, it's guarded heavily."

Nymphaea's voice rang out from the stairs.

"They are after our life force! Get them!"

"Follow me!," Hypatia moved further into the cave.

The cave wall illuminated between beats of light, revealing a sparkling blackness. Something about the twinkling rhythm reminded me of Nymphaea's dress.

We approached a ledge that revealed normal-sized dots dangling from the walls. Their axes rhythmically clanked against the rock in beat with the light. Each one looked exhausted.

"What are they mining?" Ellis whispered.

"They are mining resources for Nymphaea. If you haven't noticed, she is excessive. Most of these rocks are gemstones that allow us to cast magic. There is enough to provide for every dot in the mountain, but they sit unused in a vault near the throne."

Hypatia whistled.

In response, one of the nearby dots climbed our way.

"How can I be of service?" They chirped.

"Take us to the bottom of the cavern. Our friends from the outside need safe passage."

"Oh, dear. From the outside, you say? They don't look like demons. Is it safe?"

"They are with me," Hypatia responded. "I can assure you they are safe."

The dot nervously bounced about, trapped in indecision.

"Alright, climb in my bucket," they instructed.

We climbed into the rusty pail in front of us. The moist dirt created an earthy scent. The rocks were porous to the touch and provided a sturdy surface to hold.

"It's best you stay quiet while I work my way down. We don't want others to hear us." We took refuge under a few of the large gemstones. The rhythmic sway of the dot's movements and the water sloshing at the bottom were meditative.

Weary and tired, I started to doze off.

"Here you go," they whispered, leveling the bucket to a nearby ledge. "I can't take you any further. They will find you as they inspect my bucket."

"Thank you," Hypatia responded. "May the multiples bless you."

We climbed out.

XXIII

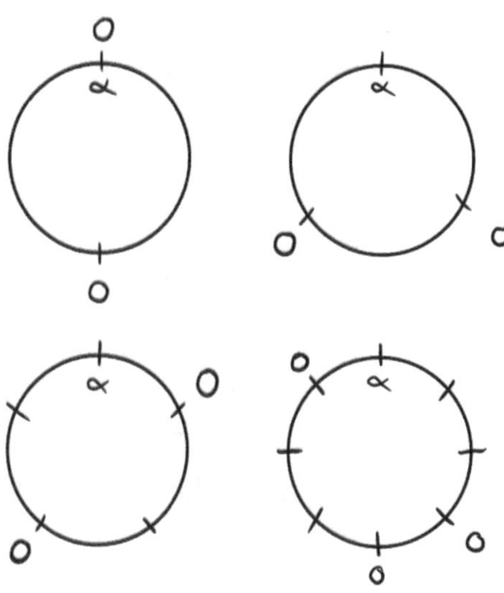

The Core of the Mountain

Eulalia

Pulses of light flowed through the tunnels and caverns, breathing life into the darkest regions. With each pass, we noticed more details. The miners covered a significant area. Their forms would block out the light as it passed through. It was unclear if there was an end to the seemingly infinite pulses grasping for the top of the mountain.

"Where are we going?" Ellis asked, breaking our silent walk.

Hypatia held up three gems.

"We need to enchant these rocks with the energy of the

gemstone so you three can get out of here." The gems radiated the orange light as it passed.

"You're taking us to the gemstone?" I said in shock.

"It's the only way out. The exit requires you to be one hundred times your current size. Otherwise, it will take you a lifetime to leave. I am not powerful enough to get you out on my own."

"Won't that cause problems for you?" Ronan asked.

"Most definitely. The prophecy says the first outsiders to visit will seek the core. If they are successful, they will release demons that will slaughter our village and burn our houses. It will take some time to convince them that is not your intention."

I grimaced at the thought of convincing the grungy crowd of anything.

"If we should fail," I said. "That prophecy may very well be true. Not because of us, but because of the curse."

Hypatia stopped.

"I do not know what plagues you, but if it is as bad as you suggest, I hope you will be successful. I will do what I can here, but things are already terrible. Maybe if we can provide others with hope it will be enough."

A warm glow flowed over an ancient village. Cracks in the rocks from each ruin lit up as the energy passed by.

The smell of moss, moisture, and earth permeated the air.

On the far side of the ruins was a cavern carved into the mountain. The scene resembled the pond atop the mountain. Instead of trees, however, ten pillars reached to the ceiling.

The source of the beating energy, the core, was inside the central platform. A large stone engraving marked the entrance. With each beat, the engraving glowed brightly. Standing just outside the pond's perimeter, I read.

"The End of Ignorance."

We each took hold of a unique stone Hypatia offered us.

"This is where we train our guards. It provides them with the power of scaling, allowing them to grow and shrink at will. Once you see the core of the mountain, touch the stone to it. It will infuse it with power. If I help, the core will not show."

She gave a bow.

"May the multiples guide you."

With a heavy breath, I stepped towards the path, which fell away. This left an inner platform surrounded by water and pillars. The water was clear.

A pillar shook to life briefly, like a beat. The vibrational hum resonated beautifully against the walls. Each of the ten pillars sang one after the other clockwise around the center. After all the pillars made a noise, there was a brief pause, and the rhythm started again. This time, the vibrating pillar did not start with the first. It had moved one to the right. Instead of every pillar vibrating, every other pillar vibrated. The next time, it started with the next pillar to the right, and every 3rd pillar vibrated. Then every

4th, and every 5th, and every sixth. Followed by a long pause. The pillars repeated.

One shook to life.
Two rang.
Three trembled.
Four gonged.
Five echoed across the chamber.

Six clanged.
Seven shivered.
Eight shook.
Nine rippled.
Ten sang.
Silence.

Two, four, six, eight, ten.

Three, six, nine.

Four and eight.

Five and ten.

Ending with six. A defining silence filled the chamber.

Each pillar split in two, redistributed itself evenly, and then repeated the pattern.

One by one
Two by two
Threes
Fours
Fives
Sixes
Sevens
Eights
Nines
Tens
Elevens
Silence.

"It's skip counting," I said aloud.

The pillars split again, as if in response to my words. Each one became five. The ringing then continued—one, two, three, four, and so on.

A word appeared in the water, glowing with the rhythm of the mountain.

"Multiples," I read aloud.

The platform leading to the center appeared.

I gingerly stepped onto the platform, testing its stability before putting my full weight on it. A tremor ran through my legs as I reached the center.

"It's beautiful!" Ellis admired and twirled to take in the scene.

Ronan seemed relatively calm as well, with a pleased smile on his lips.

All 100 pillars stood tall, evenly spaced around the circular room, singing their multiples.

One, two, three, four, five... Fifty, Fifty-one, Fifty-two, Fifty-three... Ninety-nine, One Hundred.

Two, Four, Six, Eight... Sixty, Sixty-two, Sixty-four, Sixty-six...

Three, Six, Nine, Twelve ...

Once in the three of us were in the center, an extra wave of light shot out from the podium. The pillars sank, disappearing beneath the water.

Our path to the edge of the circle disappeared yet again.

A post jutted out, marking the top of the circle. A ball rested on top of the pole with the engraved symbol α.

Ellis tapped the engraved symbol.

"Alpha," she murmured, almost to herself.

A pillar rose directly in front of α, with the engraving of a squirrel glowing brightly. The squirrel pillar moved in a circular motion around the platform. Which revealed a second pillar with an engraving of an ember.

We watched squirrel and ember move around the room. The two pillars were in a dance. Moving away from each other, then close together, then away. The rhythmic pattern formed by the moving waters below created a meditative sound. Calming, peaceful, and elegant.

For every revolution made by ember, squirrel would make two passes.

Squirrel and ember came to rest in front of α.

A new engraving was displayed on the platform.

"How many times does ember pass α?" Ellis read.

The platforms moved again, but ember lowered into the water, out of sight.

Squirrel was the only visible pillar.

"If we cannot see the pillar, how can we count the number of times it passes?" Ronan grunted.

"We don't need to see it," Ellis responded. "We only need to count every other pass squirrel makes."

Squirrel made its first pass by α.

Ronan tightened his jaw and clenched his hands into fists, showing his frustration.

"Squirrel moves at half the speed ember does," I said. "Which means squirrel needs to make two passes for every one time ember makes a pass."

Squirrel made its second pass.

"One," we counted together.

Even a single moving pillar produced a meditative, rhythmic motion.

"Two," we said in unison.

The pillar went around twice more and came to rest.

"Ember would have gone around three times," I said confidently.

squirrel lowered into the water, and Ember rose.

The engraving on the platform now read, "How many times will squirrel cross α?"

The pillar moved, revealing a third pillar with an engraving of a rat.

"I think I get it now," Ronan said, "We just need to count by twos."

"Two... Four... Six..."

"That's one pass for rat," Ellis mentioned.

Ember then lowered out of sight.

"What? It just lowered in the middle like that!" Ronan said, outraged.

"We just need to count by sixes now," I said. "That's Twelve... Eighteen..."

"Twenty-four," we all said together.

The pillar came to rest.

"Squirrel would have made twenty-four passes." We said aloud.

Five new posts, like α, rose evenly around the platform, making six poles in total.

The only pillar that remained was rat.

"How many times will squirrel cross α?" remained engraved on the platform.

"Six... Twelve... Eighteen... Twenty-four..." the pillar stopped behind the third pole.

We waited for a few moments.

Ellis and I shared a glance.

"Is it broken?" I asked. "It stopped partway through."

"Six posts... So, squirrel moved Twenty-four times around plus two," Ellis said.

"You're right! That means twenty-six times around," we said together.

The pond returned to its normal position. The

platform we used to get here rose back into place. From the ground of the center platform, a new pillar rose to reveal a beating gemstone. The beat of the stone provided the light for the cavern. The core of the mountain was within reach.

We held out the gems in our hands toward the heart.

"Thieves!" voices echoed against the walls.

"They are outsiders trying to steal our life force! Stop them!" Nymphaea's voice rang out from the dark.

Arrows flew in our direction.

We took cover behind the gemstone's pillar.

The rough surface of the rock beneath felt sharp against my skin. I danced on my knees as I tried to glimpse what awaited us.

An arrow connected with the core gemstone and sent streams of light scattered across the room as the gemstone fell to the ground next to us.

"No!" Hypatia shrieked.

Uncertain of what else to do, I picked it up. The warmth of the artifact spread through my hand, a pulsing rhythm that matched my racing heartbeat. Each beat brought new insights.

Ronan, Ellis, and I were 100 times smaller than our typical height. The massive cavern could easily hold ten thousand of us at our full height.

While clutching the artifact, I placed a hand on Ronan and my fist on Ellis. I concentrated on returning to our usual size.

In an instant, we grew twice our size and could no

longer conceal ourselves behind the podium. Three times larger. Arrows bounced off Ronan's skin. We continued to grow until we returned to our regular sizes.

Nymphaea grew rapidly.

"Get them!" she shouted.

Hypatia's wrist gemstone glowed as it grew to our size as well.

She took hold of my hand and said, "This way!"

The heart provided a dim light that enabled us to see a few steps in front of us.

Hypatia guided us around a corner and into a small opening where we had to crawl.

"In case we get separated, follow this tunnel. It will lead you out." She said.

The guard crawled into the room with us, holding a sword high.

Instinctively, I placed my hand on their chest and shrunk them until they were no larger than my knee. I placed a small pebble in the crack and let it grow until it completely blocked their access to that part of the cave.

Traveling by the light of the heart was a slow progress. We needed to wait between beats to move and could only move a few feet at a time.

Ellis tripped and slammed face-first into a rock. Blood flowed from the wound. "Let's keep moving. We can fix it later." She demanded.

Ronan got stuck for a few moments between two rocks. To get him out, I had to shrink the surrounding rocks.

The light of the exit was in sight.

"Thank you for rescuing us," I said and paused before giving Hypatia the artifact.

"Thank you." She said with a bow, but refused to take it. "Please keep it. There is no coming back from this. Nymphaea controls the narrative. She will just place it in her vault and say you took it. At least this way, it may prevent something worse."

We stepped from the dark cave into the sun's light.

"This is where I leave you," Hypatia said. "I know little of what exists beyond this cave."

"Very well." I bowed.

We touched the gemstones to the artifact to empower them and gave them to Hypatia.

"We will return the artifact as soon as we can."

"Perhaps another day," Hypatia said, entering the cave again. "For now, survive."

XXIV

Arithma's Fury

Eulalia

Our ride, Sigrid the owl, was nowhere in sight. We followed the trail of dirt and rocks until the gravel gave way to small shrubs.

Ellis stopped.

Ronan gasped.

Soot covered the few remaining bushes. Gray and black dust replaced the field of green.

We walked through the ash and soot until we came to the familiar field of Countable Grove.

I sank to my knees, my hands clawing at the soot-covered earth. The blackened remains of the innkeeper's tree split down the center but stood like a grotesque skeleton, its charred branches clawing at the sky. My chest tightened as tears stung my eyes.

Ronan's muffled sobs broke through the silence, and I turned to see him clutching Ellis's hand, both of them trembling.

Memories flooded my mind of a distant past where I helped Arithma fulfill her strategic plans. The burned landscape still haunted my deepest nightmares. It was a warm summer day, and I stood before a miniature model of the surrounding area. A group of irrationals were venturing out of the shadows.

Arithma was concerned about the safety of Origin Tower. Initially, we suffered many losses in the battle. We threw most of our resources at them. They destroyed them all. That's when Arithma had a devious plan. Her orders were decisive and effective.

She burned the forest and used its flames to slaughter any unsuspecting survivors.

"If we can't have it, then no one will." She hissed, each word seething with spite.

After our victory, I walked across the devastating damage carried out by Arithma's commands. The smoke carried with it a reminder of the catastrophic power Arithma held. I never expected to be on the receiving end of her fury.

Why? I thought. *Everything I've built. Gone in an instant.*

Feeling nauseous, I placed my hands on the ground for

stability. My mouth tasted sour mixed mixed with a bitter flavor from the ash.

Ellis gently rested her hand on my shoulder.

"Well, figure out what happened to them." She encouraged.

"You don't know, Arithma," I said, pulling her into a hug.

Ronan approached to my right. "There's no evidence of their deaths."

I squeezed Ellis tighter. The weight of the evidence felt too heavy, pressing down on me. I didn't want to admit there might be a reason for hope. I couldn't recover if they were wrong. Despite my resistance, Ronan's words caught hold.

He's right. I thought. With my vision blurred, I ran through the grove.

No bodies or bones...

"Footprints over here!" Ellis yelled. "A small group of dots came through recently."

A series of thunderous thuds that resembled loud footsteps interrupted our processing.

I turned to see a gnasher barreling towards us.

Ellis ran for the remnants of the innkeeper's tree. The carved-out hollows provided a place where we could hide.

I grabbed Ronan's arm and followed her.

Another gnasher was close behind, followed by three more.

We reached the safety of the tree.

The gnashers' heavy footsteps shook the ground as they searched for survivors.

Screams filled the air as they uncovered a dot hiding in the soot.

"Eighty-three," one of them sang joyously.

I couldn't get a good look before they disappeared into the satchel.

We remained tucked away until the gnashers ran past.

"I think it's safe," Ronan said, peering out of the hollow.

A warm, furry creature brushed my arm. I shrieked and turned.

Fear filled the creature's eyes. Its heart beat fast, and its breath was quick.

"Wriggles?" I said. "Is that you?" Its eyes softened as he nestled into my chest. "Oh, buddy. I'm glad to see you are alright." I scratched his chin, and his eyes filled with joy.

Ellis extended a hand to join in petting the squirrel. We pet him until he was running around the interior of the hollow again.

Ronan continued to search the surrounding area.

If you're alive, I thought. *The rest of them could be alive as well.*

"I figured it out!" Ronan declared.

Ellis and I ran out to meet him.

"Figured out what?" Ellis asked.

"Before we entered the crypt, we saw the gnashers camp outside Decimal City. At the Additioneer's fortress, Opal told us the story of dots going missing. Her story suggests Arithma needs the dots. They must have burned the trees to allow gnashers unbridled access to the forest. Arithma's using them to capture thousands more."

Ellis interrupted. "How does she get them back from the gnashers once they capture them?"

Ronan's face flushed with disappointment.

"That could be," I said. "If the beast is alive, they would need to feed it."

I shuttered at the thought.

Wriggles scampered out and whimpered, pleading that we stay with him.

"It's a long way to Decimal City," Ellis said, looking at the footsteps leading in that direction.

"If only we had an owl to ride," I groaned

Ronan laughed. "I hope to never ride on one of those again. I was relieved to find the owl was gone."

Wriggles pressed into my hand. I turned to him and smiled.

"You want to come with us?" I asked.

He bounced around excitedly.

His size allowed for one of us to ride. To ensure the rest of us fit, I placed a hand on his side and doubled his size. It was challenging to keep him still while we climbed onto his back. His fur provided a sturdy grip as he ran.

"Follow the gnasher footprints," I instructed the animal.

Wriggle's fur smelled of a strange mixture of nuts and ash. It was coarse and itchy. Finding a comfortable way to sit required frequent readjustment. The edges of my cloak blew around in the wind. We eventually found the rhythm of the squirrel's movements as he ran along the burned forest floor.

In all directions, we saw charred ground. A few birds flew overhead or the occasional mouse sifting through the ash for something to eat.

Decimal City started as a shiny speck in the distance and grew to the size of my fist, then my torso.

"Look there," Ellis said, pointing to a line of trees. "Not everything burned down."

Every step provided me with a little more hope.

The massive wall of Decimal City became clear. Wriggles dropped us off near a patch of forest. I shrunk him down to his normal size. Happy with the new forest, he ran into the trees.

The entrance to Decimal City was still full of trees and shrubs. On the left side, a clear burned section spread over the land. We followed the tree line until we came across the gnasher's camp.

"Let's get a better look," Ellis said, running for a nearby bush that remained unaffected by the fire.

A towering, dark gate with three guards marked a back way entrance to the city. The door opened, and three gnashers rumbled through. They walked down to the valley below.

"Are those satchels?" Ellis asked, pointing to the lumps of brown fabric scattered across the valley.

Ronan squinted a bit to get a better view. He grunted

with acknowledgment.

A serpentine path from the city led down to the camp below. While the bottom would offer us concealment from the guards, we were in full view during our descent.

"We need a way around the guards without being seen," I noted.

As if in response, the large gate creaked open. An officer stepped out and slapped a guard on the side of the head. He motioned for them to follow. The three guards' shoulders drooped, and they followed. With a soft click, the door pulled closed behind them.

"Now's our chance," Ronan said, stepping out from behind the bush.

The three of us sprinted down the path. The downward flow projected us forward. I lost my footing and tumbled. My ribs connected with a rock as pain seared through my body.

"Ah!" I shrieked.

Ronan and Ellis clumsily stopped and waddled back up the hill to help me to my feet.

"Go!" I demanded. "I'll be fine."

"You sure?" Ellis asked.

I got to my feet and limped down the hill.

"Yes! Go!" I urged.

The last turn to safety came into view as we heard the heavy doors creak open again.

I limped as quickly as I could.

Ronan and Ellis looked back up the hill, motioning for

me to hurry.

They each supported me under their arm and carried me off.

The guards above were laughing about something.

"Those green-vined monsters are so stupid," One said.

"I can't believe they fell for that." Another chided.

I made it around the bend and collapsed to catch my breath. A delicious aroma of berries and tea—the unmistakable scent of the gnasher's camp—drew me in.

We made it.

Still catching my breath, Ellis grabbed a few berry leaves from her pouch and ripped them apart in the dirt below.

"What are you doing?" Ronan asked.

"It's something Lyra taught me in the fortress. She showed me with mushroom stalks, but I think it would work with berry leaves as well."

Blood was flowing down my leg from a wound—I hadn't noticed it until that moment.

She wet the dirt and leaves and then worked the mixture with her hands.

"This should heal your leg," she said, spreading the mixture across.

A searing pain shot through my leg and up my body. My ribs hurt so much I didn't even notice the large bleeding gash in my leg.

Once fully covered in mud, Ellis etched a rune into the substance. The specks of green leaves glowed dimly as the pain subsided.

Shocked, I brushed the mud off to see the wound had completely healed.

"Do you know if it will heal ribs as well?" I asked.

"We can try," Ellis suggested. "I will need to resupply first."

I stood up and winced at the rib pain, but I could walk normally again.

"I will be slower, but should still be able to keep up. Thanks!"

Finely crafted bags covered the field before us.

Cautiously, the three of us looked around for any signs of gnashers.

The bags were empty and unattended. We walked to the nearest fabric mount.

"Anyone in there," Ronan shouted.

Ellis slapped his arm.

"Quiet down!" She said disapprovingly.

A voice inside the fabric shouted, "Yes, we are here."

"Sit tight." I said, "We will get you out."

"The core," Ellis demanded. "Use the core to grow as large as a gnasher."

"Of course," I said, pulling the artifact from my pocket. The twisting movement sent a shrill pain across my body. I dropped the artifact to the ground.

"I've got it from here," Ellis said, gesturing for me to stand back.

Once Ronan and I were a safe distance apart, Ellis grew

until we were no taller than halfway up her calf. She pushed the bag to its side and held it open.

Hundreds of dots flowed out.

"It's The Caretaker," Archy shouted to the dots inside.

Cheers echoed across the cliffside.

Ronan fearfully jumped to get their attention.

"Quiet!" he said sternly. "We aren't out of danger yet."

I stared in disbelief. Uncertain if it was safe to allow my heart the satisfaction of them being alive. *Was this a fleeting moment? How can I be sure this isn't part of Arithma's plans?* Yet, here they were. Alive and cheerful.

"Thanks for rescuing us again," the dots from Countable Grove said.

They all clumped around me, and we watched as Ellis released bag after bag.

Ronan struggled to manage the noise levels.

I peeked around the corner towards Decimal City.

The guards were starting down the path.

"We need to get out of here," I said, almost to myself.

How? Guards block our exit this way. If they alert Decimal City, we may need to fight off all of their defenses to get away.

I turned to note the gnashers on the far side of the field running towards Ellis—her back turned to them. I pushed through the crowd, waving my arms for her attention.

Nothing.

"Ellis!" I shouted, "Behind you!"

A panic rang out over the now thousands of dots.

"Remain calm!" Ronan shouted.

"Remember your training." Archy encouraged.

Ellis heard me and turned just in time to dodge one of the gnasher's heavy blows. She kicked it aside to avoid contact with the satchel and rolled across the ground.

Sigrid, the additioneer, ran out of the satchel Ellis had just released. She dodged the groping gnasher's hands.

"Additioneers, to me!" Sigrid yelled.

Within a few moments, 22 dots were surrounding her. They worked together to confuse the brute and cause them to stumble backward.

Ellis regained control and took advantage of the Additioneer's distraction as she frantically released more bags.

An angry gnasher charged her, knocking her to the ground. Their fall shook the earth, causing all the dots in the area to stumble.

Ronan and Archy were quick to their feet as they joined me.

"We need to get them into the forest for cover," Archy said.

I peered around the bend.

"We have these guards to deal with," I said, gesturing around the corner.

"There are thousands of us. They don't stand a chance," Ronan said dismissively.

"If they notify others in Decimal City, we might not stand a chance."

Sigrid and the 22 other dots were performing the prime ceremony.

Ellis, with a determined look, reached out, and with a swift motion, shrank a beast down to our size. Her disheveled hair stuck to her sweaty forehead. Exhaustion was evident in her movements as the other two gnashers charged towards her.

"Charge the guards," I said, turning to Ronan. "I need to help Ellis."

"I can help her," He insisted.

"No, after you surprise the guards and secure an escape, get the dots to the forest. I will help Ellis and then take the artifact to Al-Khwarizmi's study."

I gave him a stern look to ensure he knew not to push back.

"Very well." He said, looking defeated.

I turned to run back towards Ellis.

Sigrid transformed into an owl and shot into the air. The owl quickly swooped, clawing at one of the gnasher's heads. The distraction was just long enough for Ellis to grab hold of the other gnasher and shrink them down.

With fierce determination, Ellis narrowed her eyes and set her jaw. She grabbed the other gnasher by the shoulder. Before she could complete the spell, she returned to her standard size.

Sigrid, in the owl form, swooped in, snatched Ellis, and placed her on the grass before me.

"You fought bravely," I said, giving her a few berries.

She handed me the heart, and I felt a surge of energy pulse through my body.

"Gain your strength, and then make sure the other dots return to the forest in front of Decimal City. Remain with them until we return from the study." I instructed.

War cries rang out behind me as the thousands of dots charged the guards and proceeded up the passage.

The owl continued to claw at the gnasher.

I gripped the core gemstone and grew. In a quick motion, I snatched one of the empty satchels and placed the small gnashers inside.

The owl distracted the third, allowing me a chance to shrink them and place them in the satchel.

Ellis, fully recovered, followed the last remaining dots up the serpentine passage. I grabbed the guards and placed them in the satchel as well.

I returned to normal and collapsed to the ground.

Sigrid landed next to me.

I watched until Ellis disappeared from sight.

"Take me to Al-Khwarizmi's study," I said.

With a mighty swoop, the owl snatched me up, its large wings carrying us higher and higher. The ground shrank below. I watched from the sky as the dots disappeared into the remaining forest.

The owl placed me gently in front of a tall cave entrance and flew back towards the forest. I let out a calming breath as I stared at the beauty of the complex sea to my right.

Lyra came out to greet me.

"I was wondering if anyone else would make it," she said lightheartedly.

"I'm glad I didn't disappoint," I said, walking to meet her.

XXV

Floor Thirteen

Opal

The gate to Decimal City disappeared. Although I knew it was coming, arriving in the city hit me with dread. This time, the atmosphere felt more ominous—as if someone had pulled back a veil, allowing me to see the city accurately. The typical merchants were ready to greet us, but their smiles felt disingenuous. A plump dot blocked our way forward to give a sales pitch and toss us our first obli.

"This way," I said, waving for Oizys and Captain Numerous to follow.

We did our best to ignore merchants as they popped

up. As it became increasingly difficult, Oizys threw a wild punch in frustration, knocking a few of them back.

The path was clear of merchants after that.

To my right was the last place I saw Mortimer enjoying a meal. A few moments before, we merged to enter the building. The memory made me smile.

Lyra pulled me aside after hearing about the experience in the Additioneer's fortress. She insisted my bond with Mortimer was unique.

"Few find their inverse in this life." She said. "Combining magic does something special for the two of you. It cancels each other out, making you nothing. It's a rare form of magic."

I never thought the creepy dot I met in the gnasher satchel would become a large part of my life.

"Names?" a vined officer asked.

We ignored him and looked up at the immense architecture. Seeing the reflective surface made me frustrated. Intentionally mysterious, to drive interest in making it inside, only to find a new mystery of making it to the top. The whole thing was an unnecessary fabrication.

A pang of guilt struck me—I could have prevented Mortimer from entering had I known better.

We split up, looking for a way through the wall I noticed from the library the last time I was here. Thick bushes and trees prevented any easy analysis of the wall. Attempting to climb them or crawl through them cut and scraped my skin.

Breath ragged and quick, a girl, one-third my size, observed me with her arms folded as if attempting to read my mind.

"Do you need something?" I asked.

She jumped as if surprised I could see her. Her clothes were old and torn.

"I wanted–" She struggled to find words. "You–"

Before she could say anything further, she sobbed loudly.

"Sorry. No one will buy my pastries." She blurted out, "I cannot pay back the obli I've acquired."

I pulled out the coins I gained from the last time I was here and tossed her the bag.

"Here you go!" I said. "I don't need them."

She stared at the bag in disbelief. Her whole demeanor changed as her face lit up.

"Come with me!" She said in a hushed tone.

We ran around the first corner of the building. To the left was a beautifully landscaped park.

Being small, she could squeeze into every bush, then advise me on navigating the shrubs from the inside.

"Stay low," the dot instructed.

We ducked behind a tree with low branches, revealing the massive interior protected by the web of branches.

"Are you looking for something?" She asked.

"Yes, I need a way to get past the wall blocking the back part of the city."

"Specifically, a tunnel," Oizys said, appearing in the

bush.

"Shhh!" I demanded, trying to show the girl I was listening.

She spent some time looking back and forth between me and Oizys, trying to decide if she should run.

"He's with me," I said.

Her hesitation took a few moments to wear off.

"There is one more around here somewhere." I insisted.

"There is a secret passage that only a few dots know about. It will be hard to ensure no guards see you when you're on the other side."

Forcing his way through the bushes, the captain used his sturdy walking stick to create an opening.

"Did you just buy that?" I asked, a bit of accusation in my voice.

"I can't help myself when I see a quality staff." He said with a chuckle. "It may prove more useful than you might think."

"There are only five guards on the other side." She instructed, "One is at the large gate, leaving the city. Three more regularly check the empty areas of the space. Whatever's on that side, they do not want any dot to see."

"Why are you helping us?" Oizys asked.

Without hesitation, she pointed at me and responded, "Because she helped me. Everyone in this city is only looking out for themselves. If we worked together, we could change things for the better. Even those who understand this idea choose to follow the order mindlessly. I am helping you because I believe in working together. It is the only way to get out of Decimal City."

She sprang for the exit and stopped briefly to say,

"I hope someday the flying animals will rescue me. I like to think they rescue dots. It gives me hope. Good luck, friends."

She was gone.

One by one, we climbed out of the brush. The tunnel was precisely where she said it would be. Tucked behind one of the park bushes, we cautiously moved down the stairs. Workers placed drains at the bottom of the tunnel to allow for rain flow.

Barrels and crates filled the other side.

We quickly hid behind some of them.

An officer rested in front of the building, wearing the typical officer uniform. Another paced back and forth from the building entrance to the gate.

"How do we get inside?" I whispered.

"I can only fight one at a time," Captain Numerous said. "Any more than that, and they will notify others of our presence."

"We just need a distraction." Oizys noted. "Hang tight!"

"Wait!" I said, trying to grab his arm and remain hidden at the same time.

Surprisingly, the guards were slow to recognize Oizys. Captain Numerous and I jumped from crate to barrel to get closer to the entrance.

"You! Stop! How did you get in here?" the guard in front of the building said.

Oizys stood comfortably in the open space.

"Um." Oizys stammered as if looking for words. Discreetly, he motioned for us to move behind his back.

My heart beat rapidly.

"Well, I just kind of stumbled across this tunnel and found myself here," He said.

"This is a restricted area. You need to leave now," commanded the guard.

"What's back here?" he asked, putting a bit of a bounce on his step.

Two guards approached from behind. "I can tell when I'm not wanted." He chuckled casually. He raised his hands and backed away. "I'll leave you alone." He insisted.

We were now standing as close as we could to the entrance of the building without detection. All we could do was watch and trust Oizys to proceed with his plan.

Oizys turned on his heel when he noticed the guards behind him.

"What if I don't want to leave?" He asked. "It's nice here. It's quiet, and no merchants."

The guards were not messing around. Irritated by his banter, they cast spells in his direction.

He slyly evaded every attack. With swift agility, he ascended a stack of wooden crates and leaped from one to the next. The clatter of his footsteps echoed through the alleyway, drawing the attention of all five guards. Their heavy steps pounded the ground as they pursued him.

"Now our chance!" I whispered.

"It appears so. Who knew Oizys was so nimble?" Captain Numerous joked.

I hoped that wouldn't be the last time I saw Oizys. As we entered, the sight of a spacious entryway greeted us. To my right, a large oak desk stood imposingly, manned by a figure hunched over some papers. Shuffling footsteps echoed through the grand entryway, each step purposeful and precise. Inhaling deeply, a mix of unknown scents lingered in the air.

"We need to look like we fit in." The captain said under his breath. "Walk with purpose. Do you know where to go?"

Emptying the other side of the building would make it resemble this space. The elevator rested at the end of the foyer.

"Follow me," I said, adjusting my posture to act the part.

I briskly across the foyer towards the elevator and pressed the button.

While we waited, I wasn't sure what to do with my hands. I placed them down to my sides for a moment. Realizing that might look suspicious, I folded them.

The captain noted my nerves. "You're doing great." He encouraged as he leaned casually on his new staff.

The elevator opened, and I eagerly jumped in, ready to get a break from pretending. This elevator differed from the one on the other side of the building. It had clear numbers, one through thirteen, arranged from smallest to largest.

"This elevator is well ordered." I cheered, pressing the button labeled thirteen.

The doors closed.

I felt awkward at my comment as the captain hadn't seen the confusing elevators on the other side.

I attempted to fill the deafening silence as the elevator rose.

"The elevator doesn't have numbers on the other side of the building. The shape of the buttons changes every time you get out." I said.

"That seems like a design flaw." He responded while inspecting the metal doors.

As we rose in the building, our weight caused the elevator to shift slightly.

"It was very frustrating," I continued. "Some dots I met claimed it was a deliberate design."

"This is an incredible contraption." He said with a smile.

The doors creaked open, revealing a room engulfed in a pungent aroma of purple tarry goop. Swamp-like the substance filled the floor and walls. Each step filled the air with a slurp, as the goop clung to our shoes. A narrow walkway guided us towards the balcony, offering a view of the floors below. Purple vines snaked over a grid that hummed with movement. In the room's center, a towering rod extended from the ground to the ceiling.

The rod housed four stones. Opal engraved with a plus

sign. Quartz with multiplication. Amethyst with a symbol meaning inverse. The last one was a prime stone with no rune.

Too scared to speak, we cautiously walked down the stairs to the right of the balcony. As we passed a dark room, seawater's crisp, salty tang was a welcome change.

I glanced inside.

Hidden in the darkest part was an unlucky group of dots chained to the wall.

A chill crawled up my spine.

"Plex!" I shrieked and ran towards him.

He dangled from the wall with just enough space to stand. Cuts and bruises crawled up his sides.

"Opal! It's so good to see you." He mustered a smile, but seemed too weak to move.

"Let's get you out of here," I said encouragingly.

The captain hit the bindings with a quick tap of his staff, and the bindings fell away.

Plex collapsed to the ground, unable to carry his weight. I pulled a few berries from my unraveling satchel and handed them to him.

"Let's free the others, too," Captain Numerous said, tapping the bindings of the two other prisoners.

"What happened to you?" I asked, giving the others a few berries.

We heard the voice from below.

"I don't care who dies! Find them all!" The voice was shrill, yet authoritative and assertive.

"Yes, ma'am!" Another voice said with a tremble.

I peered over the balcony while attempting to remain out of sight.

Four ominous-looking dots with cloaks fully drawn from head to toe.

The leader had scars across her face, each covered in a purple substance resembling the checkered mass in the center. She held a bowl of beans in one hand and a staff in the other. There was an elevator ding, and then a group of 10 irrationals entered the room, chained and subdued.

Five officers prodded them forward.

"I need hundreds more. Go now! Find me more!" she demanded, giving them no acknowledgment of their success.

Unwilling to argue, they sprang to action immediately.

The leader placed a bean on a platform and muttered a spell.

The 10 new irrationals in the room shrieked as the mass in the center crawled out towards them. One by one, they were each snagged in its web.

Once complete, a familiar wave shot out from the center of the rod.

The noise from the grove. I thought. *I found it!*

As the vines retreated to the center, they brought partialed dots ensnared in their tentacle-like limbs. They wrapped themselves in a new grid layer, completing the spell.

"Tia! Aleph!" I gasped.

Before I could run down the stairs, Captain Numerous was ready. He moved swiftly to bind my arms and legs so I couldn't move.

"Now is not the time." He said, keeping his voice low. "We will save your friends, but not now!"

I knew he was right, but I was so close. I could almost feel them in the room.

Tears stung my eyes.

The elevator dinged from the floor below us.

A guard stepped out, holding Oizys wrapped in green vines.

"It appears we have some guests!" The leader cackled.

She did not order her guards to grab us.

Thinking of the heroic distraction Oizys provided, I found some courage.

"Let him go!" I demanded, still bound.

"Opal, no!" Plex said.

Arithma laughed maniacally. Unsurprised, she turned to face me. "You think you're in a position to make demands?"

The mass on the floor shifted, causing the ground to shake. The grid bubbled like boiling tar and morphed into a dragon beast, teeth bared, ready for a fight.

Unafraid, I stared into its eyes, which were at my level.

"It follows my orders, dear." Arithma said. "I used to be much prettier than I am now. These cuts are part of the job these days. Horrible things, really. They never seem to heal properly."

I gazed at the towering three-story creature, its form a chaotic mix of irrationals and rationals, packed tightly together. A guard in a robe leaned in to whisper to Arithma, who then shifted her gaze towards the window. The room filled with the hushed rustling of robes.

With Arithma no longer looking at the beast, it radiated turbulent energy.

A shockwave knocked a few guards back into the wall.

Plex and the other dots we freed transformed rapidly between their complex states.

A shock wave hit Oizys in the chest, splitting him instantly into three miniature versions of himself—each one one-third their original size. His bindings, however, remained normal. He was free.

In the commotion, we made our way to the bottom of the stairs.

"No!" Arithma yelled, looking at the beast again.

The waves of energy stopped.

Arithma fully focused on the beast as we ran for the elevator.

One of Oizys was the first to the elevator and frantically pressed the button. The other two contributed to the chaos by slipping through the grip of the guards.

The dragon clawed and scraped at the officers as Arithma fought to get it under control again.

Captain used the staff like an extension of his arm. Each officer that charged us was quickly knocked out with an audible crack to their head.

The three prisoners, Captain Numerous, three Oizys, and myself, crammed into the elevator.

Plex collapsed to the ground, his bruises looking worse than before.

"That could have gone better," Plex said. "But thanks for the rescue."

The other four dots that were with him struggled to remain standing.

Oizys stood on top of himself to prop one of them up to make space.

"We aren't out yet." The third Oizys reminded Plex.

We led them out of the building with one of us under each complex dot's arm.

We were about to reach the tunnel when a massive explosion on the upper floor sent debris raining down.

A monstrous roar echoed across the city, causing us to fall over.

Hovering in the air was the dragon, and Arithma rested in a saddle on its head.

"Citizens of Decimal City! My loyal subjects!" She chided, voice echoing across the space. "Today is a special day. It is the day we destroy disorder and chaos forever. Today, I will save you from all threats in the shadows beyond. But first, today is the day you will prove your loyalty to me!"

I wasn't able to see the other side of the decimal city. We could only watch as the horror took place.

Arithma continued, "To showcase how foolish it would

be to disobey me, I have provided you with a little demonstration. The building below me is nothing to my powers."

A jet of energy surged from the beast, causing the complete building to collapse into rubble.

We crawled away from the crumbling rubble, enveloped by a dust cloud. My hand instinctively covered my mouth as I shrieked, fearing there were still dots trapped inside.

A flock of owls swooped in and saved us from the tumbling building. Within moments, we were high in the air, overlooking the inhabitants of Decimal City.

Every dot in the city bowed in response.

The Cursed Dragon, under the control of Arithma, was a powerful force.

Dux and Phi burst through the eastern wall, blasting nearby officers across the city. Dux threw a few rocks that increased in size until they were boulders rolling toward those who pursued them.

In the chaos, dots registered freedom, and thousands rushed for the hole in the wall.

A paralyzing roar echoed from the dragon.

"Repair the wall!" Arithma ordered.

Phi fighting alongside Dux was an incredible sight to behold. The two of them created an untouchable fortress. They let the dots through that wanted to get out, but nothing could penetrate.

The beast sprang forward and landed hard on the force

field.

A slash of its claws created a visible gash in Dux's side.

Dux collapsed to the floor.

Phi ran to his side, picked him up, and threw something at Arithma. As it flew towards her, it grew in size and hit her when it was the size of a small building. Knocking her off her mount.

Arithma lay motionless on the ground.

The beast, unaffected by the blow, unleashed a surge of chaotic energy. Waves radiated out from its wings.

Some dots were partialed, and others became gnashers.

On the other side of the wall, thousands of gnashers collected the dots that escaped the city.

Arithma woke and attempted to restrain the beast.

It was just enough time for Phi to carry Dux into the forest, hopefully to safety.

In an instant, Arithma soared in our direction. Her dragon was too fast for our owls.

Despite our attempts to lose her, she remained persistent.

The dragon's claws slashed a devastating blow across the back of the owl carrying me.

My stomach lurched as we plummeted towards the ground. The world faded into the background. I tensed my muscles and closed my eyes, preparing for impact. The still-conscious owl pulled us onto its belly and rolled to its back.

We hit a patch of trees and fell through the branches

and bushes. A few scraped my arms and legs.

My body slammed into the ground, knocking me out.

XXVI

Escape

Opal

Disoriented, I struggled to open my eyes. The smell of dirt mixed with blood distracted my senses. Each movement sent shots of pain from my head to my toes. Leaves, dirt, and grass blocked my view. I wiped my brow to reveal glistening blood on my fingertips—the source of the pain.

"Help!" I tried.

My voice labored.

No response.

I felt every movement. The muscles in my face screamed in protest at my frown. Hot tears streamed down my face, blurring my vision. I leveraged the ground to turn sideways. Agonizing pain shot through my body, stealing my breath.

The owl's form rested next to me. It breathed quickly and chirped in pain. A large gash was across its back—an unnatural purple substance oozed from the wound.

Captain Numerous remained unconscious on the other side of the bird.

"Anyone there?" I said as loud as I could.

There was no response.

I listened helplessly to the owl chirping periodically in pain as it fluttered about. A new terror filled my bones as I glimpsed the owl's eyes. They shifted from a bloodshot red to a maniacal purple. Electricity pulsed across its body. A plague that was eerily similar to The Cursed Dragon.

Between sobs, I force my body to crawl to the captain. Each movement sent a chilling pain down my spine. I willed myself to keep moving and forced my body to continue despite its apparent protests.

"Captain? Can you hear me?" I tried, "Wake up!"

As I continued, the bird fell silent.

I was too nervous to look. There was no way the bird would make it out alive. I thought of the 23 dots dancing that transformed into the owl. Tears flowed faster down my cheeks. My whole body vibrated as I sobbed on the forest floor.

A rustling behind me interrupted my sobs.

As I abruptly twisted, searing pain exploded through

me.

The bird was gone.

I was alone with Captain Numerous.

A fresh wave of adrenaline took hold. The pain faded into the background. My eyes darted from bush to bush, trying to identify where the bird went.

An Oizys stepped out from behind a tree where the bird used to lay.

"Opal, you're alive," he said, running to me.

"That was quite the fall," another Oizys said, stepping into the same clearing.

Relief flooded my senses. I collapsed. Someone else was here to take over. The fear of the bird faded into the background, leaving me with the full effect of the pain.

Each Oizys looked like a smaller version of the regular-sized Oizys. They each had the same mannerisms, quirks, and movements. It was like two exact copies of him.

Mortimer and Plex stepped out from behind the tree as well.

"You scared me!" Plex said in a relieved tone.

"I thought you were a goner for sure," Mortimer stated with a glowing transparent tear falling down his face.

The third Oizys collected some berries from a nearby vine.

"Eat this," He instructed.

Each bite filled my body with new aether. My headache faded to a dim annoyance. I could move my arms and legs again.

"Thanks," I said with a mouth full of berries.

Captain Numerous jostled behind me. He groaned in pain as he attempted to sit up.

"You're alive!" I said louder than I intended and threw my arms around him.

"Ouch!"

"Sorry!" I pulled away quickly. "So sorry," I said, no longer touching him but unsure what to do with my hands. We gave him some berries, and I explained what I saw, or more accurately—what I didn't see.

"We best get moving," the captain responded, looking nervously around. "Any idea where Phi ended up? Did they make it out alive?"

An Oizys responded, "Dux is badly hurt. I don't think he will make it. They found a safe place to rest and gain strength. Phi sent us to find you. We still have a few owls to get us around if needed, but given Arithma's flying beast, it may be best to stay out of the air."

"Let's catch up with them." Captain Numerous picked himself off the ground.

He extended his hand towards me to help me stand up.

I accepted and found myself on my feet with ease. My body swayed back and forth momentarily while I caught my balance.

The three Oizys guided us back to where Phi and Dux rested.

Dux lay heavily bandaged, their head resting on a rock. He had a heavy, labored breath.

"I can't take him anywhere else," Phi stated as we stepped out from behind the nearest bush. "He's not likely to make it out of this."

Two healthy owls slept on the low branches of nearby trees.

Captain Numerous approached the bedside. "Hey, old friend. Hang in there." There was no response from Dux. He turned to Phi. "Are you able to complete the ceremony to unite the domains?"

Phi shook his head. "Dux didn't share the details. I have pieces, but I don't know exactly what needs to happen. My sister Quint might know more. I think we need to focus on getting to the study."

Dux's hand settled on the captain's shoulder, a comforting weight in the tense silence. The unexpected movement caused the group to jump.

"Listen carefully," Dux said, pulling the captain closer to his lips. "The entrance is a cave on the south cliff side of Decimal City. Arithma has cursed the land, and true order needs to be restored. It may be a millennium before dots have another chance like this. You must succeed. Promise me!"

"I give you my word," Phi said.

"Very well. My time has come."

We fondly gazed upon his form as he smiled at each of us.

His last words were, "Farewell, my friends. May the days to come provide greater comfort than days past."

His eyes closed, and his breathing stopped.

Phi fell into a sobbing heap over Dux's body.

Each one of the forms of Oizys had tears flowing down their cheeks.

Mortimer and Plex came to my side and offered a shoulder.

A shrill sound interrupted our sobs. I looked up to see Dag's infected owl diving straight at us. The purple infection dominated its eyes.

The two owls protecting the space sprang to action. Throwing off its dive.

The three bounced off the ground near us and then took flight.

"Come this way! Quickly!" Captain Numerous said.

The group was lethargic and struggled to get moving.

One of the Oizys forms grabbed my hand, pulling me forward.

I pulled on Plex and Mortimer.

Another of Oizys's forms was attempting to comfort Phi while simultaneously pushing him to safety.

The third was working with the captain to find the location of the birds and a safe place to run.

The bird circled back, and lightning cackled off its wings. It made a thunderous noise as a lightning strike shot out of its beak, connecting with the surrounding trees.

Flames erupted in the forest.

The infected bird landed on the ground beside us, and electric arcs flowed from its body and connected with nearby trees.

The other two birds tackled the bird's body to the ground and pinned it.

"Leave me!" the captain said. "I will tame this beast. This is my legacy."

Eyes determined, he turned from us to face the bird.

"Dag!" He said in a calming tone. "Hear me now. Please, friend. I know you're still in there."

Electricity sparked, and the birds collapsed next to him. Lightning shot from the owl's beak. The Captain dodged just in time.

Oizys and Plex were both pulling at my arm.

"We need to go." Mortimer said, "He will be alright. We cannot afford to lose you, too."

I helplessly glanced at Dux's dead corpse.

"Goodbye," I muttered, then followed behind Phi.

My mind was swimming in the events—drowning. Too much was happening all at once. It was too hard to process. The journey was a blur.

Unable to do anything else, I focused on Phi's black cloak. I hoped the others were behind us, but I was too scared to turn around and see. It felt like anytime I looked behind, someone was getting hurt or dying.

Just one more step. I repeated it in my mind.

Time slowed, stopped. I tripped on something. I curled

into a ball as I rolled forward, hearing muffled voices above me.

Plex pulled me to my feet. He said something and grabbed my hand to guide me through the forest. Flames and smoke consumed everything.

Plex helped me regain sight of Phi's cloak.

My muscles screamed in pain.

"Rest. I need rest." I muttered.

Plex responded in muffled tones.

We kept moving.

I think Plex would have dragged my body across the ground if my legs stopped moving.

Then suddenly, all at once, time returned to normal.

Phi had led us to a cliff side with Decimal City above it. I looked out over the steady, meditative motion of the complex sea. Waves flowed in and out. I tried to match my breath to the rhythm. The Caretaker and Lyra came out of a crack in the cliff.

A thunderous screech that cackled with electricity announced the corrupted owl was close. We all climbed inside as the third and final partial of Oizys made it just in time.

Talons clawed and scrapped at the cliff-side around us as we were just out of reach.

"Safe for now," Phi said. "Gather your strength. The Study of Al-Khwarizmi, and the last leg of our journey, awaits us inside."

XXVII

Al-Khwarizmi's Study

Opal

I awoke disoriented, unsure of how much time had passed—far too little, it seemed. My body ached with pain in every movement.

"This should loosen your stiff joints." The Caretaker said, handing me a cup made from a leaf.

It was too dark to tell the contents.

I took a sip. The dirt flavor mixed with a hint of berry was unmistakable—tea. Even with everything aching, I still didn't like it. I thought about Tia laughing heartedly the first time I tried it. I smiled. At least I'm getting better at holding it down. I closed my eyes and could smell a hint of

Wriggle's fur, nuts mixed with tea. As much as I missed Tia, knowing I was no longer alone was a comfort.

The Caretaker sat next to me.

"Opal," she said.

I turned to see her eyes fixed with care.

"I think you should be the representative of the natural domain."

She offered me the Ishango knife. The bone had notches cut into it like a tally system. The knife was durable, sharp, and ready to use.

"I don't think I'm right for this," I said timidly. "Doesn't it make the most sense for someone with the title of The Caretaker to handle this?"

"You can call me Eulalia." She said.

I couldn't believe my ears. Her words were sincere. She was talking to me as an equal, someone like her. I haven't saved thousands of dots or built a civilization to seek refuge from an authoritarian. All I managed was to lose those closest to me. Though we did rescue Mortimer and Plex, I didn't do it alone.

Eulalia continued, "I've given it a lot of thought. Lyra mentioned how you and Mortimer are unique. Mortimer is the only dot that can represent the complex domain. I think you and he were destined to be councilors. We simply wouldn't be here without the two of you. Please take it."

She held the knife closer to me.

I stared into its berry-stained hilt and extended my

hand. A pulse of energy flowed from the knife to my opal necklace. With the knife came a heightened sense of numbers. I could see values around me.

Twelve rocks kept the fire at bay. Five fingers on Eulalia's hand. Twenty-nine stitches in the hem of my clothes. However, some numbers were still too large to comprehend without counting. The number of rocks in the cave. The number of hairs on Eulalia's head.

"Are you sure?" I asked.

"I am certain." She responded with confidence. She had a glimmer in her eye that provided a deep sincerity.

"Oh, I almost forgot." She said, half standing. "Do you know how you are unaffected by the wave from the tower?"

She sat down again.

"No, I assumed some magic was involved, but haven't pieced it together yet."

"Pull out the necklace I gave you." She smiled warmly while simultaneously pulling out her own.

I pulled it out and looked at the beautiful craftsmanship. They cut the gemstone into three unique pieces, engraving the numbers 0, 1, and -1 on it. Her necklace matched my own.

"Ronan and Ellis both have one of these as well. They are called identity gemstones. It doesn't prevent all magic attacks, but it prevents becoming partialed during the binding spell Arithma uses. I would have used them for all dots in Countable Grove, but I didn't think Arithma would cast a spell on such a large scale."

"Identity stones?" I asked.

"Yes, magic applied to an identity has no effect. For example, if you duplicate something by nothing, you have the original amount. Or if we scale by one, that item remains the same size. This is because of identity elements."

"Oh," I said. "But why would one cast magic to duplicate by nothing?"

"Yes, that would be silly. But in this case, you're being protected because the magic applied to you is acting like duplicating by nothing. It's like a shield."

As the flames flickered, their warm glow illuminated the runes etched into the jewel's surface. *Such a simple concept.* I thought. *This is what's protected me?* A nagging question still lingered on the edge of my mind.

"Wait, is this why I can hear and see the wave and no one else can?" I asked.

"Not quite." She responded with a heavy breath. "I cannot see or hear the wave, either. That ability comes from a rare form of magic. The day I found you, the spell was already done."

I nervously moved my fingers over the ridges of the opal necklace.

"Arithma used aether from your body to breathe life into the beast. Powerful magic like that infuses nearby gemstones. Now the beast feeds on aether of dots it consumes. Your necklace serves a second purpose. It heightens your aether sense to detect when the beast feeds."

"You mean I know when it is eating?" I said with a confused look. "That's weird."

"Yeah, you can detect when it's eating." Eulalia chuckled. "This connection to the beast protects as well. Which is another reason I think it is best you represent the natural domain."

"Best we be on our way," Phi said, putting an end to my conversation with Eulalia.

The company prepared to embark.

"Reserve your energy," Phi instructed. "This is the deepest cavern we've found on the island. It will take some effort to get to the study. Lyra will be our guide. Mortimer provides a bit of light as well. He can follow in the back."

We all fell into a single file. Lyra was in front of Quint—Phi's sister. I was in front of The Caretaker, Phi, in front of her. Each Oizys walked between us energetically.

Initially, each step was stiff, as if my joints were rusty hinges. Eventually, the movement felt comfortable. The journey was relatively uneventful. The dim lighting meant there wasn't much to see in the cavern. We could occasionally hear the drip drop of water or get a burst of cold or hot. For most of the journey, we traveled in silence.

The chill of the air provided my body with a healthy supply of goosebumps. My teeth chattered together as I shivered uncontrollably.

"Here, you can take my cloak," Eulalia said.

"Thanks," I muttered between clenched teeth.

Even with the warmth of the cloak, it took some time for my teeth to stop chattering. It was unclear why others didn't have shivering bodies with teeth chattering together.

Oizys bounced with each step to keep warm.

I guessed we were all cold, just dealing with it differently.

The sound of rushing water filled the cave, starting as a gentle hum and growing into a deafening roar. Sunlight streamed in from the left, illuminating the path with each step, accompanied by a crescendo of noise. The air was heavy with the scent of damp moss.

"Almost there," Lyra said, stepping briefly out of sight.

I followed, stepping into a room with natural light streaming in from the waterfall above. In front of us was a thin strip of walkway and a long drop to either side. I couldn't see the bottom. Icicles clung off the platform.

"Careful on this path. It's extremely slippery." Lyra said as they started across.

I placed my foot on the thin strip and found myself collapsed on top of two Oizys.

"Easy!" they said in unison.

"Sorry, I didn't realize it was this slippery," I responded. "Thanks for saving me."

The three of us held hands as we cautiously shimmied across the chasm.

Once we reached the solid ground on the other side, I let out a sigh of relief, feeling the earth beneath my feet. Before us was a large, intricately designed building. Four marble pillars stood tall. The center of the two middle pillars was a large opening. Above the opening, an engraving said.

Al-Khwarizmi's Study
All Dots Are Welcome

The group shuffled towards the entrance, exhausted.

Once Lyra stepped inside, the fireplace ignited, offering a new warmth to the space. Lyra's glow dimmed until she was nearly see-through.

Books lined the walls. The air hung heavy with the scent of aged paper and the pungent tang of dried glue. Tia's entire collection would easily fit on a small corner of these shelves.

The scent of old books hung in the air as readers chose from various chairs and tables, promising a comfortable spot for hours of reading. On the far right side, there was a workstation set up to make books. It had parchment and dull metallic pieces.

Each of us collapsed on pieces of furniture around the study. Dust scattered across the room, causing each Oizys to cough.

I huddled close to the fire until I was warm enough to return the cloak to Eulalia.

"Wow," Mortimer said, looking at the intricate geometric patterns engraved on the ceiling.

"I've never seen a study like this," Eulalia said, admiring

more interweaving patterns on a wooden shelf.

The warm glow of light provided a comforting ambiance.

"I haven't felt this much like myself in nearly a decade," Quint said, studying a book near the fire. Her scars caught the orange light, softening them a bit. She seemed more approachable.

I got a little closer to see what she was reading. She smelled of cooked mushrooms and something sweet. The book had a bunch of symbols I didn't recognize.

After a few moments of rest, we followed Phi to the basement. The musty cave smell returned. The floor and walls were stone, as if someone had carved the room directly from the rock. As with the study above, a warm fire tracing the center of the walls provided light.

The floor displayed an engraved circle and pentagonal star. Smaller circles marked the points of each star.

Quint, still studying the book as she walked, gasped.

"What is it?" Oizys responded, ready to spring into action.

"Oh, sorry. It's this passage." She said, preparing to read it aloud. "I am the last Al-Khwarizmi tribe member."

The words reverberated off the basement walls.

"Gaia, fearful of the unknown power of our tribe, has killed each member. I am all that remains."

Eulalia's eyes darted to the ground.

"I record this hoping one day, a group of dots may be successful in completing our work. Arithmetic is a way to unite the five domains. It allows each the ability to exist in their true form while providing needs for each member. An order among the chaos."

I asked, "How can something include both order and chaos?"

Eulalia was quick to respond, "Sometimes we mistake order for control. Order, in this case, refers to understanding. If we can understand each other, it will provide a natural order. One that doesn't require our involvement. Allowing things to exist in their natural state rather than force them into something that doesn't align with their true nature."

"You mean like let a berry remain a berry rather than trying to force it into a gemstone?" Plex asked while scratching his head.

"Something like that." She responded, "Arithma learned from Gaia, and she is determined to make every dot on the island exactly like her—natural-born. We saw the crypt's negative effects firsthand. Quint and Phi are being forced to live a rational life. It creates a new form of chaos hidden under the illusion of control."

The room fell silent as we considered Eulalia's words.

Quint continued to read aloud, "Gaia's name strikes fear across all domains. She has successfully driven back the irrational tribes and some of the rational tribes. She intends to only allow naturals to exist on the island and eliminate any that oppose her. I fear she will be successful. I cannot risk her finding the study. She knows I am still alive. Because of this, I am leaving. I hope that one day, dots from the five domains will unite under the order of Arithmetic. The chamber below holds the spell to get started. If you are reading this, I hope the spell may serve you well."

The last words echoed around the room.

Lyra broke the silence, "So, Arithmetic provides understanding for an otherwise chaotic system?"

Phi was bubbling with excitement as she made connections.

"It's not a way to manipulate or control, but a way to understand true nature." She said.

Quint noted some hanging tapestries among the three of the four walls. She carefully studied each one. "I know what to do," she said.

She walked to the top of the circle.

"I need to warn you. These creatures are behind this door." She pointed to a tapestry that illustrated three monsters crawling out of the massive door.

The thought of the monsters being alive sent a chill up my spine.

"To prove themselves, the new guardians will need to fight them. You will need to work together across the domains to balance the land. If they cannot defeat these beasts, there is no hope for them as an Order of Arithmetic."

The group looked nervously at each other.

Quint did not give any pause before continuing.

"Mortimer and Plex, you represent the complex domain. Do you have water from the complex sea?" She asked.

Plex pulled out a small vial of liquid and held it up.

"Phi gave it to me once we got here." He responded.

"Excellent. Please stand on the star closest to me and face the center."

Mortimer gave me a nervous glance.

I returned an encouraging nod.

"Phi, as you are not officiating, you will need to stand in for the irrational domain. Please take your position with the square recovered from the Pythagorean Brotherhood."

She bowed in response and stood next to Plex with the square in her hand.

"Oizys," Quint said.

One of the three stood at attention, another jumped at their name, surprised to hear it, and the third, sitting on the stairs, stood abruptly.

"You were partialed and yet, remain whole. This is a rare event in our predicament. You will stand in as a representative of the rational domain. Collect the multiplicative inverse artifact known as the core of the mountain from Eulalia and stand next to my sister in the circle."

The Oizys, near the steps, confidently walked to Eulalia, and the other two nervously made their way to the circle.

"Lyra, you are the representative from the negative

domain. Take the additive inverse artifact and stand next to Oizys."

Lyra was in place quickly.

"Eulalia will represent the natural domain. Please stand in the remaining circle."

She smiled and looked at me.

"I respectfully decline. Opal has the artifact, and she will need to stand in for the natural domain."

My face grew warm from embarrassment. I looked around, and each dot smiled at me encouragingly.

"Come on then," Plex said, encouragingly.

I timidly pulled out the knife and joined the circle next to Lyra and Mortimer.

"Eulalia," Quint instructed. "You will need to wait for us in the study. I'm uncertain about the side effects of having additional dots in the room. Please wait for us there. If we do not come out within a few hours, the worst has likely taken place."

Eulalia bowed out of the room.

The five of us glanced around nervously.

Quint spoke as if she was reading something, but it was unclear what she read from, "We have our councilors. Do not underestimate the center of the circle. It is empty for a grand purpose. Nothing is difficult to specify, but it is crucial for our spell. Even though nothing is in the center, imagine one more member representing zero."

The circle's center radiated a teal glow, showing the spell was beginning.

"Zero is the missing element that unites us all. Without zero, there would be no way to unite the five domains. A gap of sorts. Naturals unite with negatives through zero, forming the integer domain."

The Ishango knife glowed brightly and floated out of my hand to the center. Lyra's additive inverse artifact also glowed and moved to the central mosaic. As each piece reached the center, a light wave flowed out and up the walls.

Quint proceeded, "Integers and fractions unite through the multiplicative inverse, the core of the mountain, or under the rational domain."

The three forms of Oizys clumped together to become whole as the multiplicative inverse artifact floated to the center, sending another wave of light up the walls.

"The real domain unites rationals and irrationals as all live on land and all are real."

The vines binding Phi's face shattered and dissolved in the air, leaving her head flowing in unpredictable patterns. She was free. A tear flowed down her cheek as the artifact joined the others in the center.

Quint's voice wavered for a moment. Her eyes filled with joyful tears.

"The real domain encompasses all dots on land, but there are still more that live beyond. The sea surrounds the entire island and unifies us under the complex domain."

The process separated Mortimer and Plex into two individual dots, and a transparent cloud connected them.

The seawater vial joined the other artifacts in the center.

Quint continued, “The complex domain is not our final unifying border, for there are likely additional domains that exist beyond that we have yet to understand. For now, we allow the fog of the unknown to be included in this domain.”

The five artifacts emitted a soft, ethereal glow, each color distinct and mesmerizing. The gentle hum filled the air, creating a soothing symphony of soft vibrations.

“These are the building blocks of the Order of Arithmetic. As councilors, you each vow to uphold the well ordering principle.”

Each artifact returned to its owner.

I held out my hand to grab the Ishango bone. As it touched my fingertips, I felt a power surge through me, stronger than before. My aching body and bones seemed to heal instantly. A greater clarity came over me. I could see larger quantities visibly. Mortimer and I shared more of a connection than I thought. The exact additive opposite. This spell would not have worked if The Caretaker stood where I was. Our bond was unique and enabled zero.

I could see Quint’s age of three hundred forty-five displayed vividly in my mind.

Lyra only recently ate the inverse fruit to convert to the negative domain.

I don’t know how I knew, but some intuition guided me, revealing the surrounding quantities.

Once each guardian held an artifact, a surge of energy

flowed from the center up the walls.

The spell was complete.

Phi, relieved to be free, sent a fresh surge of power. Grass grew at the base of her feet with a complementary set of purple flowers.

The number of petals 2—3—5—8—11 showed the flowers growth and progress.

Mortimer and Plex could change freely between each of their forms now.

Oizys could split into partials on command and then combine into a single dot again.

The chamber rumbled, and the doors beyond blew open.

XXVIII

Monsters From Beyond

Opal

Coiled like a spring, we waited for something to burst out from beyond. The room shook, knocking loose dirt and rocks from cracks in the walls. The flames around the room danced to the vibrations. Something large lurked beyond.

A sudden rush of anxiety caused me to dance nervously in place.

Lyra and Mortimer gave me an anxious look as the beast revealed itself.

It climbed out of the corner of the doorway and around the room.

Thirty, forty, fifty legs. It continued to crawl along the wall, slowly encompassing us.

One hundred one hundred twenty, one hundred forty. The artifact revealed every quantity to me. Five hundred legs in total across fifty body sections. Two eyes. Two Antennas. I didn't need to count or consider what was there. I intuitively knew.

As the creature closed in, a light pulse came from Mortimer, pushing the beast back and creating a force field around us.

"I'm not sure how long this will hold," Mortimer responded as the shield pulsated.

"Lyra, could you reduce the legs and body so it isn't so long?" Phi suggested.

"There are five hundred legs connected to fifty body segments," I added.

After a little concentration, Lyra reduced the legs and segments to just a single body. It was now harmless.

In a quick burst, the form disintegrated, leaving a small pile of dust.

Our victory was short as spiders crawled from all edges of the doorway.

Creeping in by the hundreds, their black forms devoured every warm surface. Ten thousand beady eyes reflected the light of the flames. Ten thousand legs ticking tiny tremors.

The tiny forms swarmed the shield.

"We could use division to reduce the number of

spiders," Plex suggested.

"There are one thousand two hundred fifty of them," I said.

"Oizys, that's divisible by five," Phi said quickly.

As if snapping out of a dream, Oizys sprang to action. Waves of orange light beat out from the core.

As the spiders came in contact with the energy, they slammed together. Each group of five became a single spider five times larger.

I shivered.

"Down to two hundred fifty," I said, stepping closer to the center of the shield.

"That's still a lot of spiders," Plex complained as one dropped from the ceiling and landed on the force field.

The thin shield of light weakened with each pounce from a spider.

Mortimer's shoulders slumped, his face pale. He was feeling the strain.

"Oizys, reduce their number by five," Phi said.

Another wave blasted out from Oizys.

The spiders slammed together in clumps of five. Each clumped group became five times larger, making them as large as my foot.

"Down to fifty," I said.

The force field broke under the weight of the spiders.

Plex caught Mortimer's fainting body and absorbed him so he could move freely.

The six of us danced in a circle, dodging the bouncing

spiders' attacks.

Another energy blast from Oizys caused the spiders to become ten times larger.

Five remained.

I jumped aside, narrowly missing one of their hairy legs. Before I had time to recover, I rolled to dodge a pincer attack.

Light glistened off their dark eyes as another pulse pulled them all to the center of the room like a magnet. The sound of their masses slamming together echoed off the walls.

One massive spider stood in front of me.

Despite its size, the spider moved with the speed and intensity of smaller spiders. It threw Plex into a wall.

The rocky surface crumbled to gravel on impact.

Phi grew vines around its legs, attempting to bind it in place. Her efforts were ineffective, as the spider would snap each one. Phi collided with Quint in a hurry to dodge a tree-sized leg.

"Oizys, shrink it back to normal size!" Lyra shouted, growing impatient.

"Oh, right!" Oizys said, hitting his head with his palm.

An orange wave of light completed the spell.

The spider shriveled back to its original size, stopped moving, disintegrating into nothingness.

Gasping for breath, I sat, glad we had a moment of respite.

The ground rumbled.

I let out a slow breath and jumped to my feet. Six of us faced the dark room, slowly backing away. A faint glow came from the other side, as if a massive negative dot lurked around the corner. It grew brighter with every thud. The ground jostled, and a large Gnasher stepped into the light, causing them to become transparent. Instead of a satchel, they held a colossal club.

I instantly recognized its components. Like a partitioned set, each dot individually stood inside its body. Thirty-six dots in total and a partial of 18/23. The gnasher was missing five twenty-thirds.

Oizys instantly split into five of him and began distracting the brute. His nimble nature allowed him to tease and remain unaffected by the beast.

The light in the room made it difficult to track the gnasher's movements.

"How are we supposed to stop this thing?" Lyra asked.

Before anyone could answer, the gnasher picked her up and threw her across the room. She flowed through Oizys and slammed into the wall on the other side of us.

"It's not whole," I said. "Its missing five twenty-thirds."

An invisible hand knocked me into Plex.

"We need to finish the incomplete spell," Phi said, diving out of the way of its club.

Quint threw a rock that flowed through its body and connected with the wall on the other side.

"It's from the negative domain," Lyra noted, pushing herself up. "Oizys, you need to partial me."

Phi grew a set of vines from the corners of the walls nearest the gnasher. It wrapped itself around the beast's club, preventing use. The brute looked confused for a moment and started punching the wall.

Oizys combined back into a whole and sprinted for Lyra. As their fingers passed through one another, Lyra split into twenty-thirds and began charging the gnasher.

Phi grew a path to the gnasher's core out of vines and branches.

The twenty-three partials of Lyra ran along the wood.

A purple wave flowed across the gnasher's body as Lyra jumped onto the beast.

Confused and startled, the gnasher wildly flailed its arms, destroying Phi's vine structure.

Nineteen Lyra's tumbled to the floor.

Mortimer dove. He rolled around, dodging the beast's stomping feet, and grabbed hold of the last part needed to restore the beast. He threw Lyra.

Lyra soared through the air until she connected with the beast's chest.

A burst of purple highlighted the partition as the gnasher's body dissolved into thirty-seven individual negative dots. Each one then disintegrated and vanished, leaving behind 5/23 versions of Lyra.

An aether glow replaced the darkness from the room beyond.

Oizys made Lyra whole again.

We stood for a moment, waiting for something to

spring out.

"Was that it?" Plex shouted into the nothingness.

Phi and Quint looked cautiously towards him.

"What a ride!" Oizys said, chuckling.

The group nervously chuckled, still uncertain if the danger had passed.

The flames flickered across Quint's bindings.

I felt empathy for her, still bound by the curse. However, I could see now that she was the square root of five. I could see that she was, in fact, more irrational than her sister Phi. Phi is a fractioned form of Quint. It's almost like Phi was once a rational that merged with Quint. Their sisterhood proves Quint is irrational.

I cautiously approached.

"I think we can free you," I said. "Phi, can you help me?"

Interwoven into the details of the vines was something I hadn't noticed before.

Phi interrupted my internal thoughts by saying,

"I see it now. A logical falsehood underlies the brotherhood's incomplete magic, causing its corruption. I see the tangled web of fear. So long as we can appeal to their emotional fear and provide it with a logical counter, the vines should disperse."

"Get this off of me!" She demanded impatiently. After seeing our shock, she collected herself and said, "Please."

With a smile, Phi said, "You've held the ability to disband the vines all along. Belief is an emotional response to things we do not understand. The brotherhood removed the logical element and replaced it with a belief. Beliefs provide no harm as long as the believer can put the belief aside once the truth emerges. However, this proves to be difficult when the reality of the belief is what we have grown accustomed to calling true. The brotherhood built their entire tribe on the belief that everything was rational. When confronted with the logical counter, they became corrupted by their belief."

"Can you free me first?" Quint said with an impatient smile. "Then we can discuss philosophy?"

Phi responded patiently, "This is part of the process. Sorry, sis. Beliefs are powerful, yes. But not as powerful as the truth. You are the Square Root of Five. Our nature as dots is to exist as the aether intended us to exist. Beliefs cloud judgment and make it difficult to untangle the web of lies. This lie—the falsehood that plagues you—is that you are rational. It's not true."

"I don't understand." She responded, a concerned look on her face.

Phi placed a hand on her shoulder.

"You are the Square Root of Two of the irrational domain."

A burst of energy shattered the vines.

Quint's face moved between shapes in unpredictable patterns. She was whole again. Tears welled up in her eyes. She gently placed a hand on her remaining scars. Her mouth fell open.

"Thank you!" She said with a smile.

With a swift movement, she spun around, instantly transforming the room into a lush garden with vibrant green grass, colorful flowers, and delicate shrubs. The sweet scent of spring blossoms wafted through the air, creating a refreshing atmosphere.

"That felt so good!" She said, turning to Phi. "I'm sorry, Phi. It was foolish of me to go to the crypt. Thank you for standing by me all these years."

She pulled Phi into a tight hug.

"Should we see what's in the beyond?" Lyra asked. "I don't think I could handle just walking away from here."

Cautiously, we stepped into the space where the monsters came.

The room was empty and surprisingly small. A cracked doorway marked the other side of the now bright room. We walked through to reveal a balcony overlooking a river of aether below.

The foundation of Al-Khwarizmi's study was at the source of the aether. The Order of Arithmetic is a foundational piece to our very existence.

XXIX

The Fight for Decimal City

Opal

Eulalia's face lit up when she saw us coming up the stairs.

"How did it go?" She asked.

"It was incredible!" Oizys shouted, leaping up the stairs to fill her in on the rest.

The seven of us made final preparations for the hike back out of the cave.

I was the first out of the study. The bite of cold hit me quickly.

Eulalia threw her cloak over my shoulders.

Phi grew a couple of vines across the narrow, slippery

path, leaving the study. They wrapped around the path so our feet had something to grip. For additional security, it included two railings.

"Thanks," I said, relieved I didn't need to shimmy across again.

Quint looked longingly back at the study. "I'm staying here." She said.

With a look of shock, Phi turned sharply.

"We can finally be together, and you want to split up?"

Quint looked at the ground. "I want to be with you, but I will always be curious. The Order of Arithmetic freed me. I need to know what else we're all capable of."

She looked intently into Phi's eyes.

Phi's pain gave way to understanding. She pulled her into a hug. "I will miss you! Don't do anything stupid like going into a dangerous temple by yourself."

Quint chuckled awkwardly, which caused the rest of us to laugh.

"Deal!" she said.

The journey out of the cave was much more straightforward. Phi provided vines for handhold's and footholds. Lyra could project light, allowing all of us to see easily.

I stepped out into the light of the morning sun. Its rays warmed my body.

On my exhale, I found the captain sprawled out on a nearby rock.

Oizys rushed to his side to inspect his wounds. He was

severely bruised and beaten, with a combination of shallow and deep cuts scattered across his arms and legs. A purple substance was infecting the wounds.

"He's alive," Oizys confirmed.

Lyra stepped in close. She placed a hand over his heart and closed her eyes. Her inverse artifact glowed dimly. The captain's arms and legs turned transparent, then returned to solid form. The cuts, bruises, and wounds were gone. He was whole again.

"That's incredible!" The Caretaker said. "I will remain with him until he wakes, then we will join you in Decimal City."

The order of Arithmetic proceeded onward until we stood in front of a wall of Decimal City. The wall left no hints or clues of the recent battle.

"Wow! That was fast," Lyra said in disbelief.

"What about the inside? Is it all the same, too?" I asked.

Mortimer and I combined into a single dot and moved through the thick wall. Everything was in its place, every brick as it was before the battle, every merchant cart restored.

The streets were quiet. Purple corruption covered each dot. None of them moved.

We returned to the group.

Phi grew a tree that carried us to the top of the wall and gently dropped us off on the other side. Protruding branches provided a simplicity to climbing out.

Our footsteps echoed against the silent walls, alerting

the infected dots.

They approached quickly.

"I've got this," Phi said, rolling her sleeves and stepping forward. She closed her eyes and took a few deep breaths. Bushes, shrubs, and trees jutted from the ground, creating a clear path to the tower. The top branches arched back towards the center of the path, making it difficult for anyone to get in or out.

"How I've missed using these powers." She said, admiration in her voice.

Phi made a horrible noise and collapsed to the ground, unconscious.

The attack was so quick it left us all confused. Our eyes darted around the forest path, searching for any sign. Each bush and tree created thick shadows, encouraging our minds to play tricks.

"What happened?" Lyra asked, kneeling next to Phi.

Blood pooled around her body.

Lyra worked quickly, trying to seal the wound.

"Look there!" Oizys yelled, pointing to a high branch producing a flicker of electricity.

"It's the infected owl from yesterday!" I responded.

"How did the bird get inside?" Oizys asked.

With a thunderous squawk, the bird sprang from the branch as electricity arched and cackled in the wind. It showcased incredible agility as it plummeted to the ground in a blur of blue and yellow.

Plex instinctively fell to the ground, narrowly avoiding

its talons.

The artifact's power gave me insights. Highlighted outlines of each individual dot nestled within the beast's border, like a partitioned set.

"There are twenty-three dots that make the owl," I told the others. "Two of them infected."

The owl dove again.

Oizys was ready. Just before the talons struck Plex, a protective shield filled the air.

The bird's body slammed hard against it and fell to the ground.

Instinctively, I ran to its side, needing to get close enough to heal the corruption. I tried to comfort the animal, just as Captain Numerous did with the bear.

"It's alright, little one." I said, trying to sound calmer than I felt. "We've got you now. Rest."

Uncertain of what the captain was actually doing, I said phrases that felt similar in hopes it was enough to soothe the bird.

"You've done well." I encouraged. The bird's wing twitched, and I jumped. "You can rest now." I placed a hand on its wing.

Lyra was next to me now. "How many?" she asked.

"Infection has taken hold of five out of twenty-three," I replied.

She closed her eyes as her hands contacted the side of the bird. Uncharged light radiated from the bird's body. The wound on the bird's back was healing. A substance resembling snow fell from the sky with an ethereal shimmer. Each flake touching the wound removed the infection, leaving only a bleeding gash too large to heal.

The bird chirped and wailed in pain. Its wings knocked me back.

I scrambled to my knees and crawled to its side. A new understanding filled my mind. I felt connected to the dots inside. My powers allowed me to see each one individually. This wasn't just an animal. Trapped inside were dots that knew the risks. Dots that, at one point, saved my life.

"Dag," I said with blurred eyes. "You did well. You lead well. Rest now."

As my hand fell on the bird's side, its wings stopped.

"Goodbye, Dag," I whispered, with tears flowing down my cheeks.

The scene of the bear filled my mind. Each dot touched the side of his fur and whispered something. I knew what they whispered now. Perhaps our ability to count developed from a need to remember those we've lost. Dag's name now carried a new meaning—memories I made with him. Although brief and unremarkable to most, he was my third loss. I wondered if at some point the number one was simply the name of a loved one that had passed.

The bear was my first loss. So, the word bear could mean one. Dux was my second loss, and now Dag is my third. Perhaps ancient tribes tallied their losses, hoping to remember. Holding onto those numbers provided a way for them to communicate with old memories and distant friends. Counting to them may have been messages from those they had lost. A way to connect, and always remember.

I knew the 22 other dots, briefly, but I knew them because I could number them.

"Phi will live," Lyra said. "She is too weak to come with us, though."

Oizys carried Phi back over the wall. Just four of us now. The weight of the journey ahead rested on my chest.

"Are we sure we can still do this?" I asked, shaking with emotion.

Lyra looked away nervously.

Mortimer and Plex gave an encouraging nod.

Oizys responded, "With how strong Arithma is becoming, I'm not sure we have another choice. If we wait, she may become too powerful for any hope of defeating her."

His words provided me with the encouragement I needed. I thought of Tia and Aleph. Still trapped inside the beast. Arithma took them from me. My determination grew.

"Right!" I said, "We end this today. Regardless of what's ahead, we press on until we are victorious."

Lyra brightened and took a calming breath.

We threw open the doors to the foyer of the building, ready for a fight. The inside was calm. Dots didn't even notice us. They continued about their day as if we weren't there.

"This is super creepy," Oizys said, waving a hand in front of the concierge's face. "Why is no one responding to us?"

I pressed the elevator button, and it dinged, signaling it was on our floor.

All dots continued their work, paying no attention to us.

We stepped inside. This time, I could discern numbers in the triangle shape. There didn't seem to be any pattern to the arrangement. It took a moment to locate the highest number.

"Hum... That's weird. There's no floor thirteen." I said.

I pressed the third button from the top, which showed the number for floor 10.

We rode in silence.

The elevator stopped. Doors opened, revealing a large, empty room surrounded by windows. A single bed rested in the center. We looked out of the window to the city below. Unlike the first floor, dots raged outside, frantically trying to break down the roots, bushes, and branches created by Phi.

"I have an idea of what's going on," Plex said. "I think Arithma is the cause of the corruption. She seeks control. She can only control everything if she is focused on them. She does not have the power to control this many, though."

He began pacing the floor with a blank expression. His transparent cloud tether connected him with Mortimer.

"When we were captured, she did horrible studies on dots like me. Dots from the Complex domain. She would watch what happened to us when she bound irrationals. The wave affects us differently. She wanted to infect us but failed to do so. When she looked away, or became overwhelmed with emotion, her beast misbehaved."

He stopped and turned to me. "Opal, do you remember what happened when Arithma wasn't looking directly at the beast?"

I tried to recall. I only remembered being so furious with her.

Plex continued, "When she is not specifically concentrating on the infected, they go dormant. That is why the dots in the foyer were unresponsive. She is so focused on capturing us out there that she fails to notice us inside."

He turned excitedly and said rather loudly, "She doesn't know we are here!"

"Or we are walking into a trap," Oizys stated.

Plex's face contorted with confusion. He bounced his head back and forth as if debating with himself.

"Ya, I guess that could also be the case." He said, dismissing his argument.

"If we cannot find a way to get to the next set of floors, it won't matter," Lyra said, studying the walls to find a way forward.

I turned away from the open window and canvassed the room. The floor was a solid piece of stone. The sleet is gray with a hint of purple scattered throughout. Ten windows were looking out over Decimal City and the forest beyond. The ceiling looked peculiar, though. Arranged in a grid pattern of twenty-five by thirty. It shifted and moved in unpredictable ways.

"I think floor thirteen is directly above us," I said.

The other three looked up and noticed the purple vines stretched out for the first time, forcing the dots into a grid. Hidden in the wall on the far side was a staircase.

We cautiously walked up the stairs. The ground shifted with each step, as if alive. A door was at the top of the stairs.

I twisted the knob and pushed it open.

Two green vined officers stood in front of the doors.

"WHERE ARE THEY?" Arithma screamed. "THOSE IDIOTS CAN'T HIDE FOREVER!"

There was an unmistakable edginess to her voice.

"Let's free the guards," I whispered to the other three. "This one is the square root of seven. That one is the square root of seventeen."

Lyra and I touched the first, and Oizys and Mortimer the second. Their vines dissolved to dust and blew away.

The guards looked at their hands in wonder.

"You freed us!" one of them whispered.

Stealthily, we moved in behind the guards, liberating each one we encountered. The last few were the guards standing near Arithma, searching the city for any sign of us.

Plex ran into the front desk, causing a mug to crash.

The sound echoed through the space.

Arithma turned to see us all, and a crooked smile grew across her face.

"How convenient." She cackled, "Here you are, right where I need you to be."

The dragon beast crawled into the room, its teeth barred. Horns protruding from its head and down its spine. Wings tucked gently to its side. Massive claws scraped the floor with each step.

The guards remaining under Arithma's control sprang to life.

The officers we freed stood ready to fight.

"You're killing everyone," I said. "Stop now, and we can show you mercy."

She laughed. Her eyes were determined. Face marred with corruption.

"I see everything." She chided. "I control everything. There is no more chaos except for you four!"

"Order is not the same thing as control," Oizys said. "Order is living together amidst the chaotic differences."

"Under my reign, feelings don't matter!" She responded with a menacing look. "Everyone will do exactly as they are told. It is the easiest way to ensure freedom."

The massive beast pounced immediately.

We scattered in all directions, frantically trying to get out of its way.

The dragon crashed through the wall behind us and slid out the window.

The irrationals we freed could move the walls and windows. They shifted them to put some distance between us and Arithma.

Arithma's guards sprang into action, attempting to counteract their magic.

With a clash of magic between irrationals, the room changed as if it couldn't decide what shape it was.

Oizys nimbly ran across walls, windows, and floor tiles.

Mortimer tripped and slid down the floor to land on a wall.

Keeping my balance required most of my focus.

"That one is the Square root of twenty-three," I told Oizys.

His hand glanced their shoulder.

The vines dissipated.

"Enough!" Arithma shouted. A surge of power flowed from her staff and knocked us back.

The room stopped rotating but balanced on a corner.

We all slid towards the center until it gave way under our weight, and we fell to the floor below.

Free the guards. I thought to myself. *That's what we need to do.*

Five were already free, and six remained.

Oizys, still quick on his feet, recovered the fastest. As if he knew my thoughts, he was already moving towards the next guard.

My intuition was gone, making it unclear what the guard was. The powers I gained no longer seemed to work.

The Ishango knife is gone! I must have dropped as the ceiling collapsed. In a frenzy, I looked for the knife on the ground near me.

A wall knocked Oizys back and pinned him. He remained unable to move.

Arithma was now standing on the beast as it hovered above us.

"Get them all!" She commanded.

The freed irrationals stretched the space between us, so Arithma's pet had to travel a greater distance to get to us.

Her guards sprang to action, preventing the stretching effect.

I desperately looked for the knife.

Shielded by waves of light, Plex advanced on another guard. His tether to Mortimer offered the same protection as if he was under the shield with him.

The light flickered off a surface—the knife's edge. Resting in front of a broken window.

I dove and snatched it.

Power surged through me again.

With the help of Mortimer and Lyra, Plex approached two irrationals, both square roots. I shouted their names. Each one became free of the curse.

Three guards remained.

The additional irrationals righted the room, pushing the beast further back.

"We need to free the beast. It's her source of power," Lyra pointed out.

As we rose, a fresh surge of energy exploded from the central rod, knocking us back. The force from the central rod blew away the walls and ceiling.

Oizys braced himself on the railing and caught me as I slid by.

Lyra and the freed irrationals blew off the top of the building.

The wave stopped, and the beast slammed into the ground. The building shook violently in response.

Mortimer, on the opposite side, remained on the floor.

With a groan and my body protesting, I stood. My view unobstructed to the dragon, I recognized the beast was merely a partitioned set, just like the infected owl. Every layer was a form of merged magic gone wrong. Each original member remained intact. His shoulder had fifty-six and 5/32 dots. His body comprised one thousand and 7/2. Its tail was ten and 4/3. Each was a unique number, and each would take an effort to undo the curse.

Mortimer and Plex fought the beast alone. It sprang towards them in a burst of power, striking hard against Plex's chest. Plex slammed into the stairwell behind him.

"You wretched dot! You will all pay for your transgressions!" Arithma screamed.

Chaotic waves shot from the beast.

Oizys was glowing and transparent and grew to five times his size.

I split into three smaller forms as the surrounding space grew. Then there were two of me, and back to my regular self.

The waves from the beast were causing complete chaos.

Mortimer and Plex remained motionless on the floor but shifted between their four states.

Oizys generated a warm orange shield to encircle the two of us.

We returned to our typical forms.

In a moment of complete clarity, I caught sight of her—*Tia.*

Trapped deep in the layers of the beast. Tia was in two regions. Aleph spread across three regions. I could see them. All of them.

Can I talk with them? I wondered, feeling connected to the beast.

"Tia! Can you hear me?" I tried. Her voice rang in my mind.

It's nice to hear a familiar voice.

I fought back tears. *I am close to freeing them.*

"Fight back!" I yelled. "You're trapped by the dragon."

I was no longer just hearing and seeing Tia. I could see all of them. The hundreds of thousands of dots trapped inside the dragon. Each one had a name, and each one had their desires. They were each a part of it. Each of them was providing the beast with its powers.

"Don't let her do this to you." I said, speaking to all of them. "Remember yourself."

The chaos was even greater. Massive waves of energy

broke free. The shield broke apart, leaving us shifting and changing again.

"How?" Arithma screamed in a shrill voice. "You will all be under my control!"

Veins protruded from her head and neck as she strained to get control of the beast. Its chaos was slowing.

"You can't control this!" I yelled. "Look around you! This isn't order. This is a unique form of chaos. You need to let it go."

"I will never let it go!" She screamed. All of her concentration was on the beast. The cuts and gnashes across her body were splintering and rupturing.

The dragon whimpered and whined. Followed by growls. It was becoming more agitated. More fierce. It let out a bone-vibrating roar and slashed the air, cutting a large gash across Arithma's chest. She collapsed.

The beast was unbridled and vengeful. Mortimer and Plex's body lay motionless, still cycling. My body shifted between three domains. Transparent, solid, partialed, then back to solid.

"We can stop this!" I shouted to Oizys.

I ran towards the beast. I didn't know exactly what to do, but I let my instincts take over.

Oizys followed.

The two of us touched the beast's legs, causing it to shriek in pain. Light shot out of the legs we touched. Blue healing powers fell from the air.

Mortimer split into smaller versions of himself and

provided the missing fractions to make the leg whole.

The healing power intensified until it felt like a blizzard. The dragon's foot gave way to fifty dots.

A mix of negatives, rationals, and irrationals fell to the ground.

The dragon stumbled, allowing Oizys to provide missing fractions to the leg I was touching.

Hundreds more fell to their freedom.

The dragon was dissolving into nothingness. It jumped back and whimpered as it licked its leg.

"Tia! Help us fight," I pleaded. "Aleph, fight back!"

Oizys and I charged, releasing one of its wings. The tangled web of rationalized irrationals broke free. The bodies lying on the ground shifted and morphed across different domains.

Working together, Oizys and I freed the other two legs.

The beast's remnants lunged at Oizys, who cried out as it grazed his leg.

I swiftly freed the dragon's tail and body amidst the chaos.

Dots flooded the surrounding floor.

What remained was a beating heart.

A powerful surge of energy erupted from the dragon's heart, blasting everything off the building. I felt a strong connection between Tia, Aleph, and myself. That connection helped me get a foothold. Waves of energy physically pushed me towards the edge of the building. Every step was harder than the last, but brought me closer to my friends.

I fought to keep my balance, pushing with all my might until my hand connected with the side of the heart. Its thunderous rhythm provided a palpable force.

An explosion erupted, sending a shock wave of energy flooding the whole of Decimal City and beyond.

It knocked me back, and I nearly flew off the building.

Oizys caught me again.

Feeling weak, I hobbled to the half dot, Tia, and the third dot, Aleph.

Please be alive, I begged. *Please don't die.* I made it to their side.

"Tia! Aleph! Are you there?" I pleaded. A few agonizing moments passed, but slowly, dots around woke up.

"Do you mind fixing me some tea, dear?" Tia said with warm eyes.

Tears flooded my face.

"What happened?" Aleph asked.

I pulled them both into a hug.

XXX

Chaos Restored

Eulalia

I stared at the wall of Decimal City, which revealed nothing of the battle happening inside.

"I'll be fine," Captain Numerous said. "Go! They need your help more than I do."

He pushed himself onto a rock and rested his head against the cliff.

"Go! I'll catch up when I can," He insisted.

"Thanks," I said, jumping to my feet.

I raced to the tree the order had created and scrambled to the other side. A thunderous explosion erupted from the top of the tower, propelling several bodies off.

"Opal" I gasped.

I tensed up, preparing for the impending collision as the deafening roar of the sound wave washed over me. Plants disintegrated. The city walls crumbled. I stood firm amidst the chaos.

Thick dust kicked up around me.

I reduced the number of particles in the air with a reduction spell to get a better view.

The liberated irrationals cast a spell to form a large magic net. Its blueish glow pierced through the remaining dust. Each figure gently landed on the net, providing them with a soft landing.

Coughing in the dust, I made my way towards the building. A familiar chill crawled up my spine. I instinctively touched the scar on my hand.

"Arithma," I said in a bitter tone. I braced myself, inhaled deeply, and then turned.

"You've come to end me, have you?" she cackled

Blood flowed from the edges of her mouth and the base of a large gash across her chest.

"I don't see how you're going to get out of this one," I said cautiously, bracing myself for something unexpected.

"It's not my destiny to survive this day," She said in a calm, unsettling voice. "What do you think I've been training you for?"

I gave a confused look.

"You haven't figured it out yet?" she said irritably. "I

chose you for a reason. I never stopped training you. If you couldn't defeat me, there would be no hope for what's coming."

Nymphaea's cruelty flashed through my mind.

"You mean you are protecting us?" I asked.

"Of course, I've been protecting you." She said with a half laugh. "Being a leader is challenging. We have to make the tough decisions."

"You were careless! No one needs to be that way!" I challenged.

"Come now! Remember your training. Surely you can piece this together. What was I training you for except what you've become?"

Her words were confusing. I felt a deep conflict arise inside me. *Had she played me this whole time? Am I really just doing what she wanted?*

"Do you think irrationals will always be on your side?" she asked with a deep hatred in her eyes. "The reason it took me so long to find a successor is because an irrational betrayed me long ago."

Our conversation was collecting observers, some of them liberated irrationals.

A deep, burning anger, like a wildfire, rose inside me.

"If you were helping me, why treat me the way you did?" I yelled. "You were so heartless. I realize *she* hurt you. That does not mean all dots are bad."

"All irrationals!" She hissed back. "Give it time. My fading protection will force you to confront my greatest foes. Be wary of who you chose to have by your side."

Her words hung in the air. The breeze carried her away as she dissolved into aether dust.

Cheers rang out among the crowd. A lifetime's worth of burden lifted off my shoulders. But her words left me unsatisfied. She spoke as if I followed her plans exactly.

Was it over? I wondered to myself. The events happened so quickly. Arithma had been a concern for so long it didn't feel like it was over. In my training at Origin Tower, Arithma said that our job is to ensure that order remains. Gaia gained her power by pushing back powerful dots to the far reaches of the island. What will happen to these dots now that she is gone? Now that nothing is preventing them from taking back their land.

Dots across real domains cheered and celebrated through the night.

I tossed and turned, Arithma's words echoing in my mind, preventing me from falling asleep. Something was coming. More terrible, more horrific.

The following day, I woke up to an argument.

"You are all the same!" The irrationals argued. "You only see us as workers..."

A rational from Decimal City interrupted,

"We just need to rebuild! Help us."

With a dismissive flick of the wrist, all the irrationals except Phi stormed off towards the complex sea.

Arguments broke out among the remaining dots.

"Hey!" I yelled.

Opal interrupted and stood above the crowd.

"If you need a place to stay, come with us. We are going back to Countable Grove. All are welcome. We will find a way to live that doesn't require taking advantage of one another."

The inhabitants of Decimal City cheered.

"The Grove is gone," I whispered to Opal once she stepped down.

"What?" Opal gasped.

"Arithma burned it down."

"I can regrow the forest." Phi encouraged. "We can find a safe place for the dots to stay."

With Oizys and Lyra by my side, we set off to gather the other dots of Countable Grove and met at the base of the massive split tree.

With no cover, the wind freely flowed through the grove. Arithma's life was one of immense destruction and suffering. The life I left behind. Regardless of Arithma's warning tone, I wasn't going back. I will never become the leader she intended me to be.

"Today is not the end," Phi said. "This grove will once again flourish. I will help it along."

Approaching the large tree, she touched its trunk.

"Did you know that trees talk to each other?" she asked. "They provide each other with nutrition and support. Not just trees, but all plant life. They communicate and help each other grow in times of need. Because of this network, I can help them regrow as before."

With a determined set to her jaw, she focused her gaze.

Green vines crawled out of her arms, completely wrapping the tree. As they climbed, flowers bloomed, and the tips of limbs grew leaves. The tree remained split in two, but full of life again. She then placed both hands on the ground.

The Order of Arithmetic each instinctively placed a hand on Phi. The green surged outwards, a living tide of lush vegetation. Towering trees surrounded all in attendance, their branches intertwined, creating a dense canopy overhead.

Animals swiftly made their way back to the grove, the air filled with the pleasing chirps and noises.

A larger squirrel charged at the crowd, causing everyone to leap aside, except for Tia.

"Wriggles, I missed you!" she said, arms held wide. The squirrel tackled her and chirped with delight.

I felt compelled to speak, to help forge a fresh start.

"Today marks a new era," I said. "Yesterday, we stopped the reign of Arithma. Today, we rebuild. We become better. Today, we commit to helping each other. We will build a new society, allowing all domains to live in the same space together. We will unite the five domains."

Cheers rang out around the crowd.

The knowledge of impending danger weighed heavily on me.

"Unprecedented challenges lie ahead." I continued, "If we work together and let natural order guide us. We will be unstoppable!"

That night, I found Tia talking to the group near the base of a tree. Opal rested in a comfortable chair Oizys made earlier.

Aleph sat on the forest floor, listening intently to every word Tia said.

The other four councilors were sipping some tea to welcome the evening.

"Eulalia!" Al said, joining me by my side. "You did it! We started this grove years ago. You got us all here!"

Ronan and Ellis sat down near the base of the tree.

I turned to Al. The burden was too significant.

"Come with me." I said, "I have some news to share."

We walked into the group. My shoulders were low.

"I have something I need to discuss with all of you," I said.

The chatter stopped.

"I have two things I need to mention. One of them is a new quest; we need to destroy The Hall of Names at Origin Tower. The second is that Arithma's death will bring greater evils to the land, and we need to protect ourselves."

"What evils?" Aleph asked, a hint of confusion.

"Gaia was Arithma's predecessor. She used questionable tactics to chase the other four domains to different parts of the land. The spread of news of Arithma's defeat, and lack of a successor will surely bring a tragic past to light."

The evening forest became quiet.

The following day, Al and the Guardians came with me to Origin Tower.

We found ourselves at the bottom of the hill that led up to the rickety structure. Compared to the walls of Decimal City, it was clear this tower would not withstand much force.

A few guards noted me and nervously straightened up. Things at the tower were progressing as usual.

"Hello, Ma'am," one said. I smiled, realizing how distant this past life felt.

I placed a hand on his arm to pull him out of solute.

"It's alright. Call me Eulalia." I encouraged with a smile. The Order of Arithmetic members gave a nod of encouragement.

Still looking apprehensive, the guard lowered his arm to his side.

"Order to have the gates opened and clear out the tower," I said.

His shock was apparent.

"Yes. Ma'am... I mean Eulalia." He turned to face the gate. And shouted to the other guards.

Almost instantly, the gates opened. The newborn dots remained starved and hopeless.

"Please help them," I instructed the order.

Each dot was quickly provided an abundance of

berries. Hundreds of healthy dots exited the gates.

"Arithma is no more!" I yelled, to the increasing numbers outside the tower. "You are free. Al will lead those who wish to join us back to Countable Grove. If you remain here, you may not survive."

A nervous silence fell over the crowd.

Al stepped forward.

"Dear friends!" He said, "This is real. You no longer need to remain here if you do not want to. We came to destroy Origin Tower. It holds curses that are causing pain and suffering."

A slow rumble of celebration filled the air.

"If you wish to come to Countable Grove, follow me," Al said, turning on his heel.

It took the rest of the day and most of the next morning to clear the names off the walls. Lyra eventually realized that applying healing magic to each stone dramatically sped up the clearing process.

Once completely clear, we collapsed the room and destroyed that part of the tower.

Opal and I navigated the broken flooring the rest of the way to Arithma's study.

"I need to know what else she might have been up to," I instructed.

The guard remained in front of Arithma's door.

"Arithma is dead," I informed.

He nervously moved out of the way. His armor clanked together as he braved the gaps to the base of the tower.

Taking a deep breath, I entered the room. She'd clearly been working on the binding spell for a considerable amount of time. Experiments on Pythagorean beans lay scattered around the room.

"She was trying to grow beans?" I said, terrified of the potential applications. "We need to get rid of these."

I turned to look out the window—a beautiful view of the island. A stunning view of the complex sea and the fog of the unknown beyond. A brown square marked what remained of Decimal City. To the east was something new—a thick, dark fog stretched out over the land.

"What is that?" Opal asked, looking out over the landscape with me.

"I think it is the fog of the unknown. I've never seen it this close in, though."

Lightning struck as it engulfed the ruins of Decimal City.

"I think we should be on our way," I said, concerned about this new power. "Let's get everyone out and destroy the tower."

We rushed to the bottom and ordered any remaining dots to leave. Phi grew some roots on the south side, and the tower collapsed. The dark clouds were closing in from behind us. It blocked the sunlight.

Vicious creatures popped up all over the land. Their heads shifted like irrationals but looked corrupt. They attacked everything in their path. Destroying trees, plants, and flowers. Wreaking havoc and death in their wake.

"Retreat to Countable Grove," I shouted. "Quickly. Get everyone to safety."

At the base of the hill, I looked back over my shoulder.

What will happen to the newborn dots? I wondered. *At least they will no longer suffer.*

The destruction of The Hall of Names destroyed the magic, preventing them from passing on.

After a terrifying run, we found ourselves within the border of Countable Grove. Ronan, Ellis, Archy, and Thorne waited to greet us.

"No time to explain." I yelled, "Get everyone to the grove's center and keep them there!"

Within moments, all the dots of Countable Grove clumped together in the center.

"Can you make a large shield over the grove?" I asked Oizys.

The five other councilors placed a hand on his shoulder. A vibration of light filled the grove, pushing back any darkness. The dark clouds flowed freely around the shield.

We fearfully watched as lightning struck all around us. Vicious creatures ripped into the earth like it was nothing. Jagged and fierce mushrooms, trees, and shrubs sprang up sporadically, replacing what was there before.

Animals sought refuge in the small grove. Several of them transformed into dots.

"Any idea how long this will last?" Ronan asked.

"Not a clue," I said.

Captain Numerous joined my side. "This is the greatest evil I've seen in my lifetime." He said. "I've only heard legends about its horrors. It seems you have a new power that provides safety from the chaos beyond. Do you mind if we join you within the safety of the Order of Arithmetic?

"Certainly," I responded.

Glowing dots crawled out of caves and walked into the refuge of Countable Grove.

"The Negative Caverns look to be affected as well. These dots will also need to join us," Lyra said.

Ellis and I went to the forge and crafted a long rod with five pendant holes. The rod, made of gold and silver, stood nearly two stories tall. I then took it to the guardians and instructed them to place each of their artifacts in the designated place.

Oizys grew to five times his size and planted it deep into the ground.

A beam of light shot from the newly crafted artifact, causing the fog above us to fade.

Sunlight peered into the grove again. The darkness remained just outside the perimeter.

Arithma always talked about the dangers that existed outside of her control. I never pictured this. I wondered for a moment if her reign was better than this. This is what she protected us against.

I turned to see three of the five domains working together to secure shelter. They supported each other. We were no longer subject to her control and manipulation.

Even in all the chaos, all dots could be their authentic selves, which makes this a better solution than what she offered.

Note From The Author

Thanks for reading! Most maths-inspired fiction books lean heavily towards rigorous mathematical understanding. They require significant knowledge of the subject to appreciate the plot. I wanted a story that would help readers connect uniquely with mathematics.

When I first started on this project, I stuck too rigorously to the subject of maths. Though this may captivate some, the book felt more like a textbook, which wasn't the right audience. To better engage the reader, I learned to adapt my storytelling approach. Mistakes were crucial to the process.

Today's mathematics focuses so much on right answers, that it is easy to forget that the mistakes are really where the magic happens. Sometimes, these mistakes can lead to revolutionary ideas. Failing to recognize how mistakes contribute to progress, hinders our ability to identify and rectify more significant errors. We think of past mathematicians as almost superhuman intelligence, but forget that they also made mistakes.

For Arithmetic Island, errors create a natural opposition to the plot in the story. An ancient belief we know today to be false could be a curse across the land, a quest awaiting brave souls to vanquish the evils within. Equipped with logic, the heroes discover the truth and build a new system more consistent and precise than the last. By allowing the errors/mistakes to guide the plot, it allows the reader to *learn* along with the story.

Another challenge with writing math fiction is the limitations within written language. There were a few concepts that simply didn't have verbs to describe them. To partial something or to break into smaller fractions (within the specific context of the book) were words I had to make up. I'm no linguist, so I had to improvise. Writing software rarely includes intuitive equation functions, using formulas requires a more robust program. Writers often refer to the rule of *show, don't tell.* There were some concepts that would require pages of words to come close to describing them. I opted to use illustrations instead or to tell instead of showing. To put it another way, I didn't follow all the usual rules of *good* writing. The rules simply didn't allow for me to tell the story as it needed to be told.

Though I hope to create captivating stories, my actual goal is to inspire other mathematicians to write fiction about mathematics. The maths-inspired fantasy and sci-fi genres need more contributors. Think of a murder mystery that has as many twists and turns as trigonometry. The typical functions may seem predictable, but as you apply function composition, the plot becomes a wildly tangled path that's somehow understandable. If you are a mathematician, and enjoy fiction, I want to read the stories you are creating.

I'm pleased with the way the story turned out, but it could have been told thousands of other ways. I feel as a society we lean heavily on a rinse and repeat method for storytelling. Mathematics is a natural curator of plot twists and turns. Its patterns are full of surprising elements, allowing a library of plots ripe for the taking. Storytelling is an optimal way to explore any mathematical concept. The key is for creative thinkers to defy norms and experiment with unconventional approaches in order to share the story. Stories of the future, I believe, will be written in mathematics.

Arithmetic Island was originally supposed to be a video game. I got so lost in the world-building of complex abstract concepts that I simply couldn't picture the plot. The idea was too much all at once. I opted to just tell a story first. Though I plan to make a video game eventually, sometimes we need a simpler way to tell the story to get the details right. I wish video games, stories, puzzles, and interactive elements exposed more of the mathematical background. Mathematics is brimming with unexplored worlds and unexpected behaviors. We simply need the storytellers to help craft the tools.

Arithmetic Island: The Cursed Dragon inspired me to take a closer look at Goldbach's Conjecture. Though today, I do not have the rigorous proof; I have a concept that I wouldn't have had if I didn't write this story. Mathematicians recognize the creativity needed to prove mathematical concepts. I simply didn't realize that in writing a fictional story it could help with mental blocks. If you're working on a proof and don't know where to go, writing a fictional story could help you figure it out.

Being a new author myself, I initially believed that authors always had a complete story in mind before writing. Though this is likely the case for some, it is not the approach I took. I had a vague idea of where I wanted the story to go, but I simply started writing with what I had. I put together a workable outline and wrote. Every word on the page helped me tell the story better. As the story progressed, I found new questions, characters, villains, and problems that arise naturally from following mathematical models and narratives. I let them tell the story.

In short, mathematics is so ingrained in our lives that its presence goes unnoticed. I want narratives that illuminate this beautiful subject's place within us. Maths connect us to one another. Stories help us express that connection. Thanks again for taking the time to read!

To learn more about my other projects or to connect with me, please visit my website at kylewhiting.com.

Glossary of Terms

Addition Magic
Combining Merging two entities into one.
Duplication Creating an identical copy of an object or entity.

Subtraction Magic
Removing Eliminating objects or obstacles from existence.
Separating Undoing merging or combination.
Healing Restorative magic that repairs damage, mends wounds, and restores entities to a previous state.

Multiplication Magic
Growth Increasing the size or magnitude of an entity.
Amplification Strengthening an existing property, making it more powerful.
Replication Large-scale duplication to create multiple copies of an object or entity.
Multiplicative Inverse Artifact An artifact in Nymphaea's Mountains, containing the power of multiplication and division.

Division Magic
Dividing Separating an entity into smaller, distinct parts.

Shrinking Reducing the size of an entity while maintaining its properties.
Partialing (To Partialize) Dividing a mathematical entity into smaller segments.

Counting Magic (Aether Sense and Number Recognition)
Aether The fundamental energy source fueling all magic on Arithmetic Island. The word could be substituted for mathematics.
The Aether's Almanac A book containing knowledge of magical spells, mathematical entities, and hidden information.
Prime Ceremony A ritual aligning an entity with prime-number-based magic.
The Birth of Dots An event at Origin Tower where new entities spawn across the land based on human mathematical thoughts.
Numerical Intuition The ability to instantly recognize numerical values without counting.
Counting Sight The ability to visually perceive numbers associated with objects.

Artifacts of Power
Additive Inverse Artifact Provides life to the Negative Caverns, embodying subtractive magic.
Multiplicative Inverse Artifact Provides life for all in

Nymphaea's Mountains, containing division magic.
The Pythagorean Triples A mathematical structure representing right-angled relationships and numerical harmony.
Opal Gemstone with a Plus Sign Enables addition magic, facilitating merging and duplication.
Opal Gemstone with an Additive Inverse Sign Enables subtraction magic, granting the ability to remove and heal.
Amethyst Gemstone with a Multiplication Symbol Enhances multiplication magic, enabling growth and large-scale replication.
Amethyst Gemstone with a Multiplicative Inverse Symbol Enhances division magic, allowing separation and precise scaling.
Prime Stone Amplifies arithmetic magic when applied to prime-numbered entities.
The Ishango Knife An ancient mathematical tool capable of inscribing formulas with reality-altering effects.
The Cursed Dragon Spell A forbidden incantation that summons a destructive dragon.
The Cursed Beans An artifact inspired by Pythagorean beliefs, known for its unpredictable magical effects.

Creatures & Beings

Additioneers Pirate-like dots using combination prime magic to rescue dots from gnashers and hostile places.
Complex Dots Dots from the Complex domain with

shifting personalities, fluctuating between real and imaginary states.

Dots Humanoid mathematical beings representing different numerical types: natural, negative, rational, irrational, and complex.

Gnasher Large creatures cursed to search for missing partials, collecting stray dots in an attempt to make themselves whole again.

Irrational Dots Dwell in the irrational domain.

Mortimer & Plex A dot that drank from the Complex Sea, gaining complex domain properties.

Negative Dots Shadowy, transparent dots created by the additive inverse, glowing naturally for cave exploration.

Rational Dots Natural dots altered by magic, found in purest form in Nymphaea's Mountains.

Naturals Dots born naturally into Arithmetic Island without external magical influence.

Merchants Dots in Decimal City facilitating commerce within the city.

Officers Dots responsible for maintaining order in Decimal City.

Dealers Oversee gambling and probability-based games in Decimal City's gambling halls.

Characters

Arithma Trained by Gaia, a wise and formidable ruler.

Eulalia The storyteller and guide, trained by Arithma.

Hypatia A fierce warrior resisting Nymphaea's rule.

Nymphaea The enigmatic ruler of the Inverse Mountains.
Opal A dot on a perilous journey to uncover hidden secrets of Arithmetic Island.
Thrall Arithma's devoted henchman, burdened by binding magic.
Al A dot rescued from Arithma's grasp, scarred by captivity.
Ellis, Ronan, Archy, Thorne The original dots rescued from the gnasher camp.
Lyra, Dux, Phi, Quint Leaders within the Negative Caverns.
Captain Numerous Leader of the Additioneers.
Sigrid, Dag, Lief Fighters and strategists of the Additioneers.
The Guards of Origin Tower & Decimal City Maintain order in their respective regions.
Alexander A dot dwelling in the Building of Poor Order.
Whist A well-connected dot providing valuable information about Decimal City.

Locations

Origin Tower The birthplace of dots.
Seven Bridges of the Lonely King A puzzle of passageways and logic challenges.
Mushroom Village Home to strange fungi, some edible, some cursed.
The Study of Al-Khwarizmi A chamber filled with ancient mathematical texts.

Aether Source The heart of magic itself.
Inverse Mountains A region controlled by Nymphaea, full of mysteries.
Spring of Multiples A sacred spring within Nymphaea's Mountains.
Fog of the Unknown A shifting mist where lost knowledge lingers.
Complex Sea The sea surrounding the island.
Additioneers Fortress Home of Captain Numerous and his crew.
Pythagorean Crypt A secretive, dangerous place guarded by the brotherhood.
Decimal City A bustling metropolis of trade, secrets, and danger.

Mathematical Concepts

If you're interested in studying further, here are some mathematical concepts included within the book.

Factorization Breaking numbers into fundamental components.
Fibonacci Sequence A pattern governing growth and harmony.
Graph Theory Studying connections and paths, crucial for the Seven Bridges puzzle.
Golden Ratio A proportion found in nature and art.
Set Theory Groupings of elements into infinite possibilities.

Addition Combining quantities to form a total sum.
Subtraction Removing one quantity from another to find the difference.
Separating Dividing a whole into distinct parts, closely tied to subtraction or partitioning.
Multiplication A mathematical operation of repeated addition, used to scale quantities.
Amplification Enhancing a value, conceptually related to multiplying a quantity to increase its effect.
Multiplicative Inverse A number which, when multiplied by a given number, yields the multiplicative identity (1).
Division Splitting a quantity into equal parts, the inverse of multiplication.
Counting Determining quantity by incrementally assigning numbers to elements.
Prime Number A natural number greater than 1 that has no positive divisors other than 1 and itself.
Numerical Intuition The innate or trained ability to perceive number values instantly.
Additive Inverse A number that, when added to a given number, results in zero.
Pythagorean Triples A set of three positive integers that satisfy the Pythagorean theorem.
Multiplicative Inverse Symbol Often denoted by a fraction or reciprocal, indicating inverse multiplication.
Arithmetic The branch of mathematics dealing with numbers and the basic operations: addition, subtraction, multiplication, and division.

Formula A concise way of expressing information symbolically, such as in equations or expressions.
Complex Refers to complex numbers, which have both real and imaginary components.
Natural Number A positive integer, typically starting from 1, used for counting.
Negative Number A number less than zero, often representing loss, decrease, or direction.
Rational Number A number that can be expressed as a ratio of two integers.
Irrational Number A number that cannot be expressed as a ratio of two integers; its decimal form is non-repeating and non-terminating.
Decimal A number expressed in the base-10 numeral system, involving place value and decimal points.
Probability A measure of the likelihood that a particular event will occur, expressed from 0 to 1.
Order In mathematics, refers to arrangement or sequence, often in terms of magnitude or structure.
Seven Bridges A reference to the Seven Bridges of Königsberg, a historical problem in graph theory and topology.
Logic The formal systematic study of principles of valid inference and reasoning, central to mathematics.
Al-Khwarizmi A historical figure after whom "algorithm" is named, foundational in algebra.
Multiple A number that can be evenly divided by another; central to multiplication.

Legal Disclaimer

This glossary contains fictional elements inspired by real mathematics and history. While mathematical concepts are presented with reasonable accuracy, artistic reinterpretations are included for storytelling purposes.

AI and the creation of this book

The increasing presence of AI in society is undeniable, and as a mathematically driven technology, it played a role in the production of this book. However, it's crucial to clarify the extent of its involvement. This book is a human creation born from years of dedicated study and personal experience.

What AI Did Not Do:

Generate the manuscript: The complete manuscript, from the initial 66,000-word first draft to the 76,390-word final draft, was written entirely by me. Attempts to use AI for story generation proved unsuccessful, as it struggled to grasp the nuances of my vision.

World building and story elements: I conceived and developed the story outline, characters, locations, and descriptions. I drew inspiration from mathematical history, concepts, and personal experiences. AI did not contribute to the unique perspective that informs this story, which is rooted in four years of college education and seven years of independent study. This involved extensive research, including reading numerous (and often dense) mathematical texts, learning to draw and illustrate, studying piano to enhance musical descriptions, and honing my writing skills—all without AI assistance.

Create the Artwork: I created all illustrations using a

digital drawing pen and tablet.

Replicate Personal Connection: AI cannot replicate my connection to mathematics or my unique worldview, shaped by my background, passions, dreams, and desires. These profoundly personal elements are essential to this work and cannot be artificially generated.

How AI Was Used:

AI tools like Gemini and ChatGPT served as supplementary resources during the writing process:

Writing Process Insights: AI provided general information about the writing process, offered advice for a first-time author, suggested ways to improve writing style, and provided practice writing prompts.

Name Generation: AI assisted in brainstorming names for characters, locations, and creatures. While AI-generated suggestions often required refinement, they proved helpful. A typical prompt I used looked like this:

"You are an English professor helping young authors create unique names. I need a name for a fierce, dictatorial ruler who rules with fear on an island made of numbers. The name should incorporate arithmetic elements and numbers and draw inspiration from Latin and ancient civilizations like the Greeks and Romans."

World building and Research: AI aided in exploring concepts, researching historical topics like the Pythagorean Brotherhood, Plimpton 322, and the Ishango Bone, and generating ideas for myths, legends, and deities.

While AI provided a convenient way to access information and explore different perspectives, it was crucial to fact-check its output due to occasional inaccuracies. My mathematical background was invaluable in this regard. Image generators were also occasionally used for visual inspiration.

Editing: ChatGPT's current paid version includes a feature known as projects. I used this feature to get my initial edits out of the way and before I looked for beta readers. This is the prompt I used in each chapter of the book.

You are a professional editor. Help me edit my novel to prepare it for publication. I will provide you with one chapter at a time. I need to know the following:

Does this chapter flow well overall? Are there major clarity or pacing issues?

Does the dialogue flow naturally and feel like a real conversation?

Are characters relatable and interesting?

Marketing and Promotion: I plan to use AI tools to assist with marketing and promotion after publication. I do not have insights on this at this time.

Personal take on AI and writing fiction:

AI is not better: I thought that once I finished my second draft, I could upload the manuscript to ChatGPT and get my final draft. I wasted precious writing time reading over chapter after chapter of AI-generated garbage. AI is flowery writing and, at first glance, might seem better—it's not. Trust your personal writing skills. The human element you bring will always be better than the machine-generated stuff. I know it's slower. Trust me, it's worth it.

An honorable mention—Had I used more AI-generated content, it would have caused me to do 2-3 times more edits, and I would have missed critical opportunities to learn more about the environment and characters.

Embrace imperfections: AI has the illusion of being perfect. If we look deeper, the truth will be easy to uncover. The human element is what we connect with. There are things in our writing that AI will never be able to understand simply because it's not human. Our society pressures us to be perfect, but I've come to learn that it's within the imperfections that true beauty resides. If your writing is good but not perfect, it would be a mistake to ask AI to improve it—leave the errors—it's proof a human wrote it.

Good software is more important than AI: I used several software tools to help produce this book. AI is unlikely ever to be able to replace the consistency or speed of these types of tools. Please keep AI out of them.

Conclusion:

AI proved to be a helpful tool for research, brainstorming, and refining certain aspects of the writing process. However, it's essential to emphasize that the core elements of this book—the narrative, characters, world, artwork, and unique perspective—are entirely my creation. AI can be a valuable resource for aspiring writers, offering support and guidance, but it cannot replace the human element of storytelling. Your story is worth telling, and AI can be a tool to help you share it with the world.

www.ingramcontent.com/pod-product-compliance
Lightning Source LLC
Chambersburg PA
CBHW030341310726
48979CB00001B/127

* 9 7 8 1 9 6 7 3 6 0 0 4 8 *